To Howard.

with all best w

for a Very Ha

from Ben + Dot Jordan

April 1979.

The Complete Book of GUNDOGS IN BRITAIN

The Complete Book of GUNDOGS IN BRITAIN

Edited by TONY JACKSON

Barrie & Jenkins London

First published 1974 by
Barrie & Jenkins Ltd,
24 Highbury Crescent, London N5 1RX

Printed and bound in Great Britain by
Cox & Wyman Ltd,
London, Fakenham and Reading

ISBN 0 214 20027 2

Contents

Introduction by Tony Jackson

The Complete Book of Gundogs in Britain was first conceived through the correspondence pages of *Shooting Times and Country Magazine*. Three years ago I wrote an article which was, basically, a plea for tolerance in the working gundog world. For far too long, I felt, the shooting scene had been dominated by labrador retrievers and English springer spaniels to the virtual exclusion of a wide variety of other species. The causes of their decline are well known—a changing pattern of agriculture, the domination of the driven shoot, the dwindling interest in walking up, particularly over pointers and setters, the acceptance that the labrador and springer certainly met most shooting mens' requirements without need to look further afield and, most significant of all, the influence of the show bench which has undoubtedly proved the near ruination of a number of erstwhile useful working breeds.

I suggested, in my article, that whilst labradors and springers—and now golden retrievers which have displayed, of late, that they are a force with which to be reckoned—might seem the obvious choice when selecting a gundog, it might be well worthwhile casting an eye over the field, for the sake both of variety and, more important, encouraging breeds which are clinging on by a whisker.

The response was startling. It rapidly became clear that not only was there tremendous interest in the several less well-known breeds of spaniel, but that pointers and setters, so long a minority group, were the cause of much speculation. Quite as significant was the discovery that throughout the country there existed pockets of owners and breeders struggling to maintain their particular breeds and all aware that, unless interest could be stimulated amongst shooting men, the breeds would inevitably vanish from the working scene.

It was obvious that the enthusiasm existed, but the knowledge and the means of obtaining information were sadly lacking. Here, then, is assembled all the facts and figures for the potential owner, for the man who would like to cast his eye over the field. Some of the runners may, perhaps, appear non-starters but there are certainly several breeds well worth a second glance.

I believe that the pattern of shooting in this country, established for the last hundred or so years, is likely to alter in the near future. There will still be driven shoots but the costs may force many tenants and owners to hunt for their birds rather than have them driven to them. If this is so there will be a resurgence of interest in the various spaniels whilst, of course, the so-called dual purpose pointers are already claiming much interest and attention.

From my point of view this has been a fascinating book to edit. I have been enormously impressed by the sheer enthusiasm displayed by the individual authors, an enthusiasm which, I feel certain, will be conveyed to the reader. All have in common a sensible approach to the gundog, whatever its breed, and a determination once again to see it flourish.

The days of 'dog breaking' are, thank goodness, long since past and the modern approach to training is based, to a great degree, on commonsense and an assessment of the dog's character as an individual. This, the psychological approach, is firmly established, something for which both shooting men and gundogs can be thankful.

It would, of course, be a feat beyond human ken to collect the views, opinions and forthright statements of some eighteen practical working gundog folk, and hope for a uniform approach. The writers are all individuals, several well-known in the field trial and test worlds for their outspoken views. I have little doubt that some of the opinions expressed in this book will create controversy. If by so doing the interests of the several breeds are furthered then the book will have served its purpose.

Tony Jackson, Editor of *Shooting Times and Country Magazine*, has always had a keen interest in gundogs and field trials, owning both labradors and English Springers—not to mention assorted terriers. His dogs have all to be practical bag-fillers, though he will readily admit that they lack the refinements found in field trial dogs.

He has a wide experience of shooting in this country, from wildfowling to game shooting and firmly believes that the shooting scene could well be enhanced by a resurgence of the less familiar gundog breeds. He believes that there is a sound case for trying to foster interest in the breeds which were once household names in the sporting world but which are now seen only on the show bench. Apart from shooting, his other interests are horses, foxhunting and hounds.

Chris Sutcliffe, a headmaster in the Midlands, first acquired an interest in labradors whilst living in Kenya in 1956–60. During that period he ran dogs in field trials and had considerable success, winning every available field trial trophy. He trained and owned the first field trial champion in East Africa—a labrador, Grizelda Girl. He has exported labradors, one coming third in the Dutch Retriever Championship and has competed in trials and tests with success.

He belongs to twenty or so field trial societies and is now instructor to the Ashbourne Dog Training Club. He learnt much about gundogs from Mr T. Greatorex and has helped Peter Moxon start many young labradors.

Mrs M. Waters, born and educated in Scotland, has had an unbroken ownership of golden retrievers for 50 years; she has been a field trials judge for 15 years and a Kennel Club judge (retrievers) for over 20 years. She picks up on local shoots several days a week from August 12 to the end of the shooting season and occasionally runs in field trials and at working tests at Game and country fairs. She used to race cars, ride in point-to-points and whipped in to hounds.

Chairman of the Flatcoated Retriever Society, **Dr. Nancy Laughton** has bred these retrievers for the past 25 years. Apart from training flatcoats, her hobbies include shooting and competing in field trials. In 1968 she published 'A Review of the Flatcoated Retriever'.

An expatriate Scot married to an Englishman, **Mrs Isabel Heywood** had a country childhood, spent partly in the south of England. She acquired an abiding interest in the countryside, including field sports and has always been an owner of, and interested in, gundogs. She saw her first curly-coat at a Game Fair and promptly obtained a puppy. She would like to see a serious revival of the curly-coat as a worker. Mrs Heywood is a member of the United Retriever and Curly Retriever Club.

Peter Moxon, one of Britain's leading professional gundog trainers, has, from his childhood in Kent, spent his entire life with dogs, establishing his first training kennels near Canterbury in 1942, then moving two years later to Herne. In 1957 he moved to the Midlands, setting up a large establishment at Beoley, near Redditch where he has remained ever since.

One of the first exponents of the 'psychological' approach to gundog training, he has influenced numerous owners and trainers, both by his willingness to assist individuals and also through his weekly articles which have appeared in *Shooting Times and Country Magazine* since 1943. His book *Gundogs: Training and Field Trials* has acquired the status of a classic and is now in its ninth edition. He has written for other magazines and broadcast on radio and television.

A familiar figure at field trials and tests throughout the country, Peter Moxon has done much to put the gundog right 'on the map' as far as the ordinary owner and shooting man is concerned.

Keith Erlandson, born in Yorkshire in 1931, has been a professional gundog trainer for the past 16 years, winning numerous awards at trials, both with springers and cockers.

Of Swedish descent, he spent ten years as a gamekeeper after leaving school, gaining a useful insight into the use of gundogs from a practical angle. It is this background which has to a great extent shaped his approach to gundog training—'I am adamant that a top field trial dog is nothing more than a super shooting dog'.

A colourful figure, his forthright views have not always proved popular, and as the reader will discover, his approach to cockers and their training is based on a commonsense, down-to-earth attitude.

Married, with four sons, he lives in Denbighshire.

Mrs Elizabeth Blest married into a country-loving, shooting family and, in her own words 'after being pulled through and over every sort of bush and wall by an excited dog' was provided with a gun and taught by Churchills. She has shot in England and Ireland and first decided to buy an Irish Water Spaniel after a holiday spent in Co. Clare. She has never regretted the decision nor been without one since.

Hal Leopard spent his childhood and youth in Wales—hence his interest in Welsh Springers, owning his first in 1928. He has bred and exhibited Welsh Springers extensively and has bred and handled the only Spaniel of any variety since the last war to win an open stake at field trials and become a Show Champion. He now breeds field trial English Springers and owns the prefix Rushbrooke.

He joined the Welsh Springer Spaniel Club in 1929, became Honorary Assistant Secretary in 1933, Honorary Secretary in 1947 and President in 1965, a position he still holds. He also founded the Welsh and English Counties Spaniel Club in 1966 to promote Spaniel field trials and has been Chairman since its inception. He is on various other committees and a member of most other Societies running Spaniel field trials. He is also on the Kennel Club 'A' list of field trial judges for Spaniels.

It was doubtless significant that **Mrs Rachel Furness** was born at Sutton Scarsdale, Chesterfield, home of the late Mr William Awkwright who maintained a strong kennel of Clumbers in addition to his perhaps more famous pointers. Mrs Furness is on the committee of the Clumber Spaniel Club, the Irish Setter Association and the Setter and Pointer Club; she is also a member of the Ladies Kennel Club. She has exported Clumber Spaniels to America, Australia, Sweden, Italy and France.

Mrs J. R. Freer has been concerned with gundogs all her life. She is devoted to country pursuits and her greatest interest above all is in Sussex Spaniels, which she has bred, and studied, for many years.

Roger Hall Jones, born at Malvern in 1946, has been a member of the Field Spaniel Society since 1963 and Hon. Secretary since 1969. He is Hon. Treasurer of the United Spaniel Society, Hon. Secretary of the Mid-Western Gundog Society and is delegate to both the Kennel Club Liason Council and the National Gundog Association. In conjunction with his mother, Mrs A. M. Jones, M.B.E., he breeds Field, Cocker, American Cocker and Sussex Spaniels.

Mrs A. M. Jones, M.B.E., was born in Cheshire and educated in Belgium. She served in the Womens Royal Air Force from 1939–46 and was awarded the M.B.E. in 1946. She took up the serious breeding and exhibiting of Cocker Spaniels in 1959, followed by Welsh Springers and Field Spaniels in 1962 and, four years later, American Cocker Spaniels. She bred the first bitch Show Champion American Cocker in this country, judged the breed at Cruft's in 1972 and is Secretary of the American Cocker Spaniel Society of Great Britain.

Dr. James Maurice has been Hon. Secretary of the English Setter Club since 1950 and has judged at field trials in England, France and Italy and attended them in America and Belgium. In the shooting field his experience has been equally widespread, having shot in Arabia, India, North Africa and Italy. He has contributed articles on English Setters to *Shooting Times and Country Magazine*.

Michael McGarry was born in Belfast and now lives in Westmorland. He has bred and trained Setters and Pointers for the past twenty-five years, though since 1959 the emphasis has been on Pointers. His Toberdoney Pointers have been consistent award winners on the bench and at Trials. He has judged shows and trials all over the British Isles, and in Europe and Japan.

Apart from his profession as a dentist, his interests are horses, Welsh ponies and Hereford cattle.

Mrs Jeanette Allan, who lives at Banchory, Kincardineshire, has spent all her life surrounded by gundogs and has bred, owned and handled champions and field trial winners. She is a founder member of the Gordon Setter Association and is the current President; she has judged Gordon Setters at Crufts. Amongst her other interests she is keen on salmon fishing, gardening and dancing.

Mrs A. Mason owned her first dog at the age of three, and she has kept Irish Setters for well over 30 years. She has always lived in the country and, naturally, has trained her dogs and used them both for rough shooting in the earlier years and later for field trials and work on grouse moors. Nearly all her dogs have obtained awards, both in Championship Shows and at field trials, while dogs of her breeding have been exported to many different parts of the world. She is also a breeder of thoroughbreds, a keen gardener and naturalist. She is secretary of the Irish Setter Association of England.

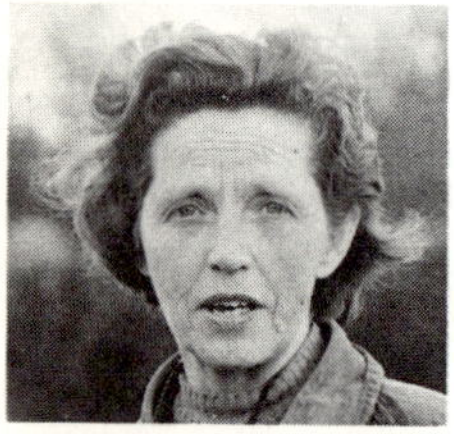

In 1960 **Mrs Louise Petrie-Hay** founded the well-known Waidman Kennels, where she breeds and trains German Shorthaired Pointers, Vizslas and Weimaraners. Since then she has made up a GSP Champion and a Weimaraner Champion, whilst a Vizsla bitch from this kennel went Best Any Variety at Crufts before the breed acquired Championship status.

A keen supporter of field trials, Mrs Petrie-Hay has won many major awards since 1962. She is on the British Kennel Club's 'A' list of judges for GSPs and is an 'A' list field trial judge for GSPs, Vizslas and Weimaraners.

Mrs Michael Gosling has always lived a country life, being surrounded by horses and dogs, enjoying all the good things of the country to the full. Her husband was Master of the East Essex Hunt for 23 years and though she had trained spaniels for a number of years, it was not until a bad accident prevented her riding again that she took up training in serious fashion. She admits that she has caught the field trial 'bug', though she obtains most enjoyment from beating and picking-up. A keen observer of dog psychology, she insists on breeding her own dogs to train, and admits that having had spaniels all her married life she has never been tempted to try other breeds.

1 Labrador Retriever

HISTORY

That the Labrador is the most popular dog in the shooting field is borne out not only by observation of any shooting party but also by the Kennel Club's registration figures over the last ten years, when this breed has had the second largest registrations *of any breed.*

The reason is not far to find, for the Labrador is kind, gentle, and adaptable, being outstanding both for his work and as a companion. In fact, so good is his temperament and trainability that most guide dogs for blind people are either pure Labradors or first crosses.

It is these qualities, plus others which have been developed, which we have received from the past and which it is our duty to improve, if possible, and pass on to our successors.

The origins of the breed are, as with most gundogs, shrouded in mystery but, as far as can be ascertained, the first Labradors as we recognise them were obtained by Col. Peter Hawker and Lord Malmesbury from fishermen plying between Poole in Dorset and Newfoundland. We do know from records of the first European settlers that the Indians of the New World did not know cattle, horses or dogs. So it must be supposed that dogs were introduced there by Europeans.

Dr. Caius, a physician in the reign of Queen Elizabeth I, wrote a 'systematic arrangement of British dogs', but there does not appear any description slightly resembling the Labrador. The earliest authentic description is given to us in Col. Hawker's *Instructions to Young Sportsmen*, written in 1814, when he describes the Labrador, or St. John's dog, and the larger Newfoundland, and goes on to say that the Labrador 'is by far the best for every kind of shooting . . . oftener black than of another colour and scarcely larger than a pointer . . . pretty deep in chest . . . has short or smooth hair, does not carry his tail so much curled . . . is extremely quick in running swimming and fighting'.

The next source of information on the early history is from Major M. Portal's *The Gun at Home and Abroad*, and Blaine's *Encyclopaedia of Rural Sports* (1852), and it appears that after the Earl of Malmesbury started his kennel, the Duke of Buccleuch founded a kennel of three obtained from the Earl. The Earl of Home also kept Labradors between 1850 and 1860, whilst the Netherby Kennel was started in 1860 by some dogs given to Sir Richard Graham, who in turn gave some to the Earl of Verulam who started a kennel at Gorhambury, so that the Labrador geographically spread from north to south.

From here onwards the history of the Labrador is on firmer ground. All field trial Labradors can be traced back by male descent to Malmesbury's Tramp, Netherby Boatswain, Malmesbury's Sweep, Ch. Ilderten Ben, Bright and Buccleuch Bachelor. It is thanks to these early breeders who kept accurate records and took infinite care to keep their strains pure, that we can trace the subsequent history of the Labrador and now reap the benefit of that care in the working abilities of our present day dogs.

Perhaps the greatest influence is Malmesbury Tramp born in 1878, who produced Avon of the Buccleuch kennels and sired B. Nith out of B. Gyp, who in turn mated to Munden Sarah produced Munden Sixty. This latter dog in Lord Knutsford's kennel proved the greatest influence, not only in his own kennel but in a great many others at that time.

Of the two names closely associated with the Labrador one is Lord Knutsford who was, when he first became interested in the breed, the Hon. Arthur Holland-Hibbart. He obtained his first Labrador from the Earl of Verulam, Munden Sybil in 1884, who can be traced back to Boatswain. Munden Sixty won a first prize at the Carlisle Dog Show in August, 1900, and when mated to Munden Scottie produced Munden Single the first Labrador ever to gain an award at field trials—a certificate of merit at the International Gundog League's trials at Sherborne near Warwick in 1904, as a five year old. A year later she gained a similar award at the I.G.L.'s trials at Sutton Scarsdale as did another Labrador, Mr Own Mansel's Satrap.

Lord Knutsford from 1884 until his death in 1935 kept unabated his tremendous interest in the breed and in his 79th year won the Non-Winners Stake of the Herts, Bucks and Beds F.T. Society, handling his own dog Munden Squeezer. His interest was in the dual purpose of the breed—looks and work—for he showed and worked his dogs, gaining awards in 1903 at the first Kennel Club show in which the breed was recognised as an entity.

The first Labrador to gain real recognition in trials was Major M. Portal's Flapper (descended from Malmesbury's Sweep) who came 2nd in the 1906 I.G.L.'s trials; winning in 1907 the Kennel Club's trials at Horstead and by winning in 1908 the I.G.L. trial became the first Labrador Field Trial Champion.

The first Labrador to win the I.G.L. Retriever Championship, inaugurated in 1909, was the Duchess of Hamilton's bitch Dungravel Phoebe (daughter of Flapper, whose career ended before the classic was staged) followed by Mr A. R. Ruxton's Hunsdon Zulu in 1910 and Capt. A. E. Butter's Peter of Faskally in 1911. This win not only brought to the fore the claims of the Labrador as a retriever but also brought into focus its ability to be trained to the highest degree, for Capt. Butter's training and handling was based on the working and handling of sheepdogs, and consisted in controlling and working the dog by signs and whistling. The huge success of his methods in handling his dogs Peter and Patron of Faskally and his wife's Percy of Faskally, aroused tremendous interest and caused many to attempt to imitate his methods, with, alas, little success. His methods started the controversy which has continued to this day of handling versus natural working ability.

Before World War I the Labrador came on to the shooting and field trial scene, demonstrating that it possessed in abundance the desirable attributes required by the shooting man, and by 1914 it was already beginning 'to dent' the Flatcoated Retriever dominance, both in trials and in registrations at the Kennel Club. Notable Labradors of that era were, in addition to the above, Capt. A. Glen Kidston's Gwendoline and Juniper, Mr T. W. Twyford's F.T. Ch. Peter of Whitmore and Ch. Type of Whitmore and Mrs Quinten Dick's (later Lorna, Countess Howe) Champions Ilderton Ben and Withington Dorando. In fact, from 1909 to 1913 in the Retriever Championship, only once did a Labrador fail to win it and that was in 1912 when Col. Weller's Flatcoat Meeru won.

During the First World War no trials were held, but they were then resumed without interruption from 1919 until 1938. However, in 1916 it was decided by interested Labrador followers to form the Labrador Retriever Club to protect and encourage the breed, with Lorna, Countess Howe, as its first secretary. This was due to a dog, Horton Max, winning a challenge certificate for Labradors, a dog who was, in fact, three-quarters flatcoat and only one quarter Labrador!

From 1919 until 1938 the second name so closely associated with the Labrador dominating both the show and working fields, was that of Lorna, Countess Howe, whose kennel prefix was Banchory. Together with her kennel manager, the late T. Gaunt, she made the greatest dual purpose kennel the breed has ever known. Her association with this breed commenced at the turn of the century when she owned a Labrador called Scandal of Glynn. This dog died at five of canine typhus and sired only one litter; a litter of twelve bitches and one dog. So much had the character of Scandal impressed the Countess that on his death she determined to obtain, if possible, a descendant of her first love. After much trouble she traced and obtained the only dog, who became Dual Ch. Banchory Bolo (the first dual champion in the breed) and firmly concentrated her overwhelming interest in Labradors.

A brief history of this dog may be of interest. Bolo came to Countess Howe as incorrigible and was offered to her with the proviso that if she couldn't do anything with him, she must have him destroyed. With kindness and perseverance she got him steady and trained to trial standard. However, when being exercised a whip crack frightened him and he bolted in terror. He returned to her Ladyship's room at 5 a.m. and on waking at six she found that he had two deep chest wounds, hind leg and back torn to the bone and a 3 in. groin tear. As the nearest vet was eight miles away, she personally put 23 stitches into him, disregarding advice to put the dog down. In the following season he became a F.T. Champion, subsequently gaining the coveted Dual Champion title.

Between the two World Wars the Labrador, both numerically and in working powers, went from strength to strength. The Kennel Club's figures for Labrador registration are:

1912: 281 registrations
1922: 916 ,,
1932: 1205 ,,

In 1925 the Yellow Labrador Retriever Club was founded with the late Major A. Wormald as its secretary. It is invidious to draw a distinction between black and yellow Labradors; indeed in 1910 a yellow, Mr F. Straker's Rust, scored the first 'yellow' win at a field trial, whilst the first 'yellow' award in a field trial was in 1908 in Northumberland when Mr Straker's Sandy gained reserve. In 1927 the first yellow to gain title of Field Trial Champion was Hayler's Defender owned by Mrs Lindsay Stewart and the first ever yellow Champion was Major Lucas's Hawesbury Joy. The zenith in this period for yellows was in 1938 when Mr R. G. Heaten's Cheveralls Amber was bracketed equal 1st in the Retriever Championship.

Before returning to the inter-war years history of the breed, a brief word on the background to the yellow Labrador is called for. The majority of yellows to-day can be traced back to a dog Neptune, bred by Major Radclyffe, of Hyde Wareham, who did a great deal to further the cause of this colour. Neptune was by Mr M. Guest's Sweep ex Nell, the latter bred by Major Radclyffe—both Sweep and Nell were bred from dogs imported here by Mr Guest and Major Radclyffe. The outstanding kennel of this colour during the 1919–39 period was Major A. and Mrs Wormald's Knaith kennel, very well known in both bench and trial circles. However, H.M. King George V successfully supported the show bench, with dogs entered under his Wolferton prefix, all however being worked first and foremost.

The outstanding dog of this period is Countess Howe's F.T. Ch. Balmuts Jock which, in the Retriever Championship alone, has created a record still standing—3rd in 1924, equal 3rd in 1925, 1st in 1926, 3rd in 1927, 1st in 1928 and 1929. The only other dog of this period whose record is at all comparable to Jock's is Lady Hill-Wood's F.T. Ch. Hiwood Chance whose Championship record is 1931, 3rd, 1932, 3rd, 1933 and 1934, 1st.

During the inter-war years Lorna, Countess Howe, had six Field Trial Champions as well as 12 Challenge Certificate winners in her kennels. However, many others espoused the Labrador cause, for a number of well known post-war supporters of the breed started their trial careers during this period e.g. Lady Hill-Wood, Lord Rank, Major H. Peacock, R. MacDonald, Mrs N. J. Heywood-Lonsdale, Mr H. A. Saunders, Mrs M. F. Radclyffe, Mr V. Routledge, etc. With the win of the Retriever Championship in 1920 by Tag of Whitmore owned by Mr T. W. Twyford, we have the only solitary instance of this event being won by three successive generations in the male line. In the early 20s inter-breds enjoyed a brief popularity and of the six inter-bred F.T. Champions, five had a Labrador sire, whilst the Labrador has produced more Dual Champions than any other breed. During this time these were: Banchory Bolo, B. Sunspeck, B. Painter, Bramshow Bob, Lochar Nessie, Titus of Whitmere and Flute of Flodden—four owned by Countess Howe.

During the 1939–45 War trials did not take place, and breeding was restricted to those key strains whose perpetuation was essential to ensure the breed's survival. Quickly the breed went from strength to strength, the Kennel Club's registration totals being:

1946: 4839 registrations
1952: 3196 ,,
1962: 8139 ,,
1972: 13,880 ,,

thus demonstrating the breed's popularity.

Lorna, Countess Howe's active participation in field trials now diminished. However, a number of others came to the fore, but those who wished to combine work and showing have decreased. Since 1946 only three Dual Champions have occured, E. Winter's Staindrop Saighdear (3rd in the 1946 Retriever Championship), Mrs A. Wormald's Knaith Banjo (both these dogs were yellow) and one black, the last to gain this title, in 1951, R. MacDonald's Rockstead Footspark. Perhaps the most successful in this sphere in the post-war era has been Dr. T. S. Acheson with the Ballyduff prefix; three who also tried hard to combine both sides were Mrs R. G. Williams (Mansergh) Mrs A. Radcliffe (Zelstone) and N. D. Robinson (Mallardhurn). Recently Mr & Mrs J. Johnson of the Reanacre prefix have supported both shows and trials but with two entirely different strains of dog.

A number of kennels of pre-war origin have continued with great success until to-day, those of Lady Hill-Wood (Hiwood) and Major H. Peacock (Greatford) perhaps being the two strongest. Lady Hill-Wood with the help of her daughter Ann has consistantly produced home-bred field trial Champions (eight in all), winning the Championship with Hiwood Dipper in 1960 and her daughter repeating the achievement in 1964 with Dacre Hiwood Frank.

Major H. Peacock has the finest 'eye' for a working dog that this country has ever known. He rarely bred but has produced seven field trial champions including three winners of the Championship. In this he has been ably assisted by Jack and Keith Chudley of the Harpersbrook kennels.

'The Chudleys', as they are known, have been involved in the production of more field trial champions than any other two people in this country. For the late Mr F. George they produced five champions, and have made up at least four others for clients, in addition to helping the 'Major' to produce his.

Lord Rank, until his death, was another who continued from the inter-wars years, with first W. Brunt and later A. K. Manners as his trainers. This 'Scotney' kennel was responsible for six champions but never won the Championship.

Mr Vincent Routledge another pre-war devotee, and a close friend of Lord Rank's, was a breeder and always ran home-bred dogs under the Hallingbury prefix, winning the Championship with Blackbird in 1962.

Mrs Radclyffe continued her husband's interest in the yellow Labrador and she too home-bred her working stock; she is perhaps the last true producer of the dual purpose dog. This 'Zelstone' kennel has produced four Champions; Z. Darter came 2nd in the 1948 Championship with Z. Moss again coming 2nd in the 1957 Championship.

Another keen trials supporter Mrs. J. Heywood-Lonsdale (Shavington) has produced over the years some excellent dogs; some bought, some bred, all home-trained, which include four Champions. Her last F.T. Champion, Shavington Jericho Jane, was at her working peak when the foot-and-mouth disease outbreak of 1967 stopped all trialling.

Last but not least of the pre-war adherents who carried on into post-war is the Staindrop kennel of the late Edgar Winter, now ably continued by his daughter Mrs J. Hayes. Eight Champions have carried this prefix, the last being Staindrop Trim.

Two dogs that have had tremendous influence on the breed are J. Annand's F.T. Ch. Glenhead Zuider (1st 1949 Championship)—almost every trial dog now has this dog one or more times in its pedigree. The other is F. Bell's F.T. Ch. Grouseadee. Both dogs had one thing in common, natural ability, one of the necessary qualities in a dog. Mr F. Bell, now in his eighties, has produced very many trialling dogs, in fact one trained by him, Peteradee, won the 1959 Retriever Championship, for Mr R. S. Wilkins. However, Mr Bell very rarely bred but used his 'eye for a dog' in purchasing likely puppies.

Perhaps the greatest mating of post-war times was the one between Greatford Pettistree Shadow and Mackland Honeysuckle. This was repeated three times, each mating producing a winner of the Championship i.e. F.T. Ch. Greatford Churchfield Jet (1951), F.T. Ch. Oxendon Dan (1953) and F.T. Ch. Greatford Teal (1955).

From 1946 until to-day, the Pinehawk Kennels under Andrew Wylie have quietly produced some very good workers, although again few have been home bred. In fact he piloted Mrs William Owen's Treveilyr Swift to victory in the 1948 Championship—this dog then went on to become the first post-war International F.T. Champion.

The late Lady Auckland (Cromlix) had an interest in the Labrador as well as in her beloved Cockers, Pointers and Setters. She made up her three Champions Skip, Dinah and Tiny, in the short space of five years, a record I think, and held the Championship a number of times at Cromlix.

In the 50s two kennels came into existence that have had great influence breeding-wise on subsequent generations, namely that of the late W. Lawrence Taylor (Galleywood) and Mrs B. Harcourt-Wood (Glenfarg). Each had an extremely able trainer and adviser who helped to ensure success. Dick Male trained and handled the Galleywood dogs, including Shot which won both the 1957 and 1958 Championships, the first dog to do this since Lady Hill-Wood's Hiwood Chance (1933 and 1934). In addition, he trained two other Champions for this kennel (Swift and Brook). Dick Male then transferred to Mr W. Charles William's Berrystead Kennel, and under his expert guidance this Kennel has become one of the most prominent in the country, having produced five Champions since 1964. In the Championship alone the Kennel won the event in 1966, and since 1965 has gained, in addition, two 2nds, one 3rd and one Reserve.

Mrs B. Harcourt-Wood's Kennel was fortunate in having G. Meldrum as its trainer. Its first Champion, Norham Blackie, won the Championship in 1956 and the subsequent record shows that Mrs Harcourt-Wood, too, had an eye for a dog and chose her foundation stock carefully, for nine dogs gained the F.T. Champion title, the last two under Mr Meldrum's name. Seven titles were gained over eight years, and the kennel records include in the Championship only, two firsts (1956 and 1963) one 2nd, one 3rd and two Reserves.

Another kennel of great note is Mr R. N. Burton's Brackenbank Kennel. Over ten years in the 50s it produced Champions in Trigger, Merry, Tanner and the greatest of all Jasper, whose Championship wins were three times 3rd and one 2nd.

Mr D. F. Cock's Roffey prefix is still going strong, starting with his first Champion, Roffey Gay Time in 1954, followed by Roffey Gamble, Dunlop and in 1971 with Brampton Dipper. The Holdgate prefix of Mrs G. M. Benson is now well known, for in 1956 her first champion was Creedypak Stella, one of Col. H. C. Kingsford Lethbridge's breeding, and was followed in 1963 with Holdgate Boffin and then Holdgate Vestra and now with Holdgate Willie whose Champion record is 1st in 1971, 3rd in both 1972 and 1973. Also from the West Midlands area is Mrs C. C. Lamb whose first Champion was Lambdale Simon and now has Lambdale Shaun, and Mrs S. Monckton whose F. T. Ch. Jericho Jingo is the culmination of much thought in breeding and work in training.

Mention must be made of Mr R. S. Wilkins Sendhurst Kennel trainer R. G. Baldwin, whose first Championship success was 1st in 1959 with Peteradee, followed by F.T. Champions Sendhurst Galleywood Monty, Sendhurst Zelstone Tinker and Sendhurst Sweep (1st 1965 Championship).

Two kennels in recent years that have made their impact on the trial scene are those of Capt. D. Mackinnon whose Beinhmhor Tide won the Championship in successive years, 1969 and 1970, ably handled by P. White, who also, in 1967, made Annan of Lanfine a Champion too. The other kennel is the Hedenhampark one of Frank Clitheroe who, together with his brother John, has produced some excellent dogs, winning the 1973 Championship with H. Holcot Fay.

Credit must go to some fine breeders of notable families of Labradors for the pre-eminent place the breed holds in field trials e.g. D. Mackenzie (Staxigoe), R. Grant (Ruro), D. I. Walker (Mycur), Lt. Col. E. Unwin (Eastwalton) and Mrs D. Purbrick (Strattonley).

Lastly, the interest, knowledge and support of H.M. The Queen with her Sandringham prefix must not be overlooked. Not only does she actively support trials by competing and giving ground, she is the Patron of the Labrador Retriever Club. In this she carries on a family tradition. Her grandfather, H.M. King George V, supported the show bench with his working yellow Labradors; her father King George VI ran Windsor Bob in trials and she first entered trials with a Hiwood dog in 1954. Since then, in the very capable hands of W. Meldrum (son of G. Meldrum, Mrs B. Harcourt-Wood's trainer and mentor) the Sandringham Kennel has gone from strength to strength.

Her Majesty's first F.T. Champion was Sandringham Ranger who gained his title in 1964, followed by Sherry of Biteabout in 1967 and Skipper in 1968. In the 1973 Championship Her Majesty had two dogs running, Sandringham Flora and Sandringham Sydney, both gaining awards with Sydney being Reserve, and, what is equally important for the future, both home-bred.

What of the future? The prospect looks bright, for in the last four years 16 'new' owners i.e. owners of dogs who have never before gained the F.T. Champion title, have dogs which gained this status. Many more men and women are training Labradors and competing in field trials, for all stakes are heavily over-subscribed and have been so for the last ten years. I would hazard a guess that even with the increased number of Labradors, there have never been so many trained dogs as at present. It must be remembered that the sole object of all field trials is to promote the training and thereby, the breeding, of suitable gundogs.

CHOICE OF PUP

'Fools breed them, wise men buy them' so goes the old saw, and not without an element of truth. For in checking the history of the breed it appears more people have successfully trialled Labradors that they have not bred, then have been trialled by owner-breeders.

I must quote my old friend Peter Moxon, who, in one of his books, wrote that for work one *must* go for pups whose parents and grandparents worked, and to reduce the chances of 'buying a pup', preferably one whose immediate ancestors had gained awards at trials. I know full well

1. W. Meldrum, handler to H.M. The Queen, with two of her dogs.

2. A typical Labrador.

the criticism that trials receive ('circus dogs' etc . . .) but until a better way is devised this is the only way we have of measuring a dog's performance in the field. We require of a gundog that it is quite steady to moving game without caution and free from chase, sits quietly at a drive, and walks well at heel, retrieves shot game tenderly to hand on order, is obedient to whistle and signal, marks well, and has natural game-finding ability. The Labrador has all these plus style and drive and, in addition, is a good companion.

It is impossible to discern at eight weeks old these desirable qualities and it is better to leave the choice, if you can, until the pups are at least 4–5 months old; even then it is not always possible to choose the best! One of my best bitches was the most backward at this age. From whom to purchase can be a problem; watch the advertisements in the sporting press and, if one sees what appears to be a likely litter write or ring up for the pedigree. Check this for award ancestry and then, if satisfactory, go to see the pups. Dog breeding is still an art and not an exact science. The two great forces which combine to form an individual are heredity and environment, so a healthy, well made bitch of normal temperament and good working ability should have been mated to a similar dog, both of which should be proven free from hip dysplasia, and hereditary cataracts. These two hereditary diseases are, unfortunately, entrenched in the Labrador ranks and only stock certified free under the BVA/KC schemes should be bred from.

There is another adage at this stage: 'Choose in haste, repent at leisure'. Decide on the sex and colour you want—bitches are a trouble six weeks in the year, a dog can be trouble all the time! A good horse is never a bad colour and neither is a dog. Yellow or black—take your choice! The only guide I know on the difference between blacks and yellows work-wise are some notes by the Royal Army Veterinary Corps on their use of Labradors for tracking in Malaya during the troubles. There they found that on the whole, blacks were more persistent when hunting a faint line. The third colour, brown or chocolate, is not very widespread and is centred in the Notts.-Derbyshire area. I have yet to see one at trials or in the shooting field, for the main interest in this colour stems from the showing fraternity.

Looks are not paramount; go for the extra-bold one, the pup that comes up to you unafraid. At any rate, do not take home a nervous one. Check by moving among the pups, watch them at play and clap your hands unexpectedly to see their reactions. Providing they have been well reared and are not temperamental almost any will do. When by itself throw your balled handkerchief a little way in full view of the puppy to see if it will run and pick it up. If it doesn't after two or three tries discard and choose again. All things being equal choose one with a hazel eye, nice short otter tail thick at the root, nicely set little ears, wide skull, dense thick coat and with a free active movement. A Labrador should be a natural retriever. I have no time for any dog that has to be 'forced' to perform its natural function.

CARE

Once at home its future life depends upon you, both as a trainer and kennelman (or maid). Its first requirements are warmth, food, shelter, exercise and affection. Whether it lives in or out is a personal matter—my E.A.F.T. Ch. Grizelda Girl lived her first three years in a four roomed flat; a number of Countess Howe's dogs lived in with her. However, it is better if the pup can have a kennel and run outside, where it can exercise itself and have a home of its own. It must never be allowed to roam at will without supervision.

Its kennel can be what you will, with the proviso that it must be dry and draught proof. Dogs can stand a great deal of cold but not wet or a draught on their backs. Gone are the old days of four meals a day with all the attendant trouble. If, like me, you have a full time job, then there are many modern foods that now make puppy rearing easy. I usually feed milk plus a few spots of cod liver oil in the morning and meat in the evening. At all times there is access to a dry feed *ad lib.* (in a bowl or hopper) and plenty of water, so that it can help itself.

There are many proprietary foods on the market in meal or pellet form. I prefer the former for it takes longer to eat and helps to keep the puppy occupied. Meat can be tinned or fresh to suit your convenience. Sheep and cattle tripes are excellent if obtainable, for the pup then has something to gnaw and chew. I also like to provide my pups with a big shin bone, for not only does it occupy them gnawing-wise but they soon learn to pick it up at the point of balance and it strengthens their neck muscles —good practice for carrying hares later on.

Exercise night and morning, even if only briefly prior to feeding, helps the pup to learn to be clean and not foul its kennel. As the puppy gets older then its evening exercise becomes longer and its meals correspondingly larger, so that at about four months old it has $\frac{1}{2}$ pint of milk daily and approx. 1 lb of meat. The meat is increased to $1\frac{1}{2}$ lb daily at about six months, and I continue milk and hopper feeding until about a year old when I stop both and increase the meat quantity, taking as my gauge the condition of my pup. By way of a change I occasionally feed meal, pellets or biscuits.

It is important that at about 12 weeks old the puppy is inoculated against distemper, hardpad, leptospiral jaundice and hepatitis, and the booster inoculation given when required by one's vet. If there are skin vermin then a suitable shampoo and/or insecticide from the vet is needed. On purchase, ascertain if the pup has been wormed for round worm. If not, do so immediately and again in about two months' time and again at eight months' old with suitable tablets from your vet.

This period, from puppyhood to adolescence, is vital to the future of your dog, for not only must its physical needs be met but it must be made wordly-wise i.e. used to people, travelling, cattle, and strange loud noises. It is important that it gets plenty of exercise on different types of terrain, for it is at this stage that it learns to cope with every type of going e.g. plough, rough pasture. Incidentally, I have found that a puppy which spots and chases birds usually becomes, with experience, an excellent marker.

TRAINING

When to commence training? This has already been done, by taking the pup out and about, teaching it its name—easily done by calling it immediately before feeding and walking on a lead. To get a puppy used to a lead merely let it run at exercise, trailing 3–4ft. of old rope. Once this is familiar, gently hold the rope for a minute or so, gradually increasing the period until it walks with you wherever you go. On no account at this stage jerk the rope hard or try to make the puppy walk on your left. Labradors naturally carry and *on no account must a puppy be scolded* for bringing its finds to you. If by chance it picks up unsuitable objects e.g. your wife's tights, *do not* chase after it but run away, calling the puppy. Usually it will run after you and when it reaches you, bend down, stroke its head and then gently remove the object by putting your fingers in its cheek between its teeth and prizing its jaw open—if the dog bites it bites on its own cheek. Lift the object upwards saying 'Dead'. Do *not* indulge in a tug-of-war.

Do not start the formal aspects of its training too soon for you cannot put an old head on young shoulders. I would suggest you start seriously at about 7–8 months old, unless the puppy is headstrong. In discussion with eminent trainers I find that the greatest error we all make in training, is an incorrect assessment of the pup's temperament; the kinder the puppy the later the start of its training. Do not commence training unless you can continue for at least two months, for if you do stop and start, the routine is never established.

The equipment you require is simple—at the outset a lead, whistle (not the pea or referee type, for a dog's hearing is sensitive and a loud one is not required), a rubber or tennis ball and two dummies. Mine are merely old socks stuffed with nylon tights and tied at the ankle end. The beauty of these is that they are light, easily dried (a shaking will do) and replaceable. Later three or four additional dummies will be required, one of which must float and one be heavy, at least 3 lb. Rabbit skin dummies are not necessary, unless at the advanced stage your dog refuses to pick fur. I find that if initially I use skin dummies the puppy is more inclined to bite and play with them. Artificial scent of any kind on a dummy is unnecessary for it soon acquires a scent easily recognisable by the dog.

If you wish you can purchase for advanced training a dummy launcher from Windrush Guns Ltd. at Witney, Oxon. A blank pistol of some kind is essential ; it can be of the starting type or the special one for dog training made by the Acvoke Co., or even a cap gun. The Acvoke one is excellent but I find it too heavy for my use.

Commands should be kept to a minimum. Mine are: to recall the dog, either its name or pip, pip, pip, pip, pip, on the whistle. To sit, either one long whistle blast, a raised hand policeman halt fashion, a foot stamp, or verbal 'hup'. In the advanced stage I introduce the dog to sitting to a raised gun. To walk to heel, the verbal 'heel' or 'ssssss . . .' To retrieve, verbal 'fetch it'; to jump, verbal 'over'; to go away farther from one, verbal 'get out'. To release retrieve, verbal 'dead'. To prevent a wrong action a verbal 'no'. To release from drop verbal 'Tch', the old waggoner's chirrup to a horse to move if close by, or other command.

In all training a quiet spot is necessary where distractions are minimal, though open fields are not essential, for much basic training can be done on your lawn. Unless the pup is headstrong I do not teach it to walk to heel until the training is well advanced, all I insist is that the dog walks on a slack lead on my left. This is done easily by manoeuvering the lead so that the puppy is on the left and occasionally giving a sharp jerk, so that the puppy realises that it is more comfortable not to pull on the lead. The lesson duration, almost until the field stage, is between 5–10 minutes daily and I usually combine evening exercise with training. I do this winter and summer, for even in the dark much useful training can be done, but throughout the watchword must be make haste slowly.

The first dummy is usually thrown when the puppy is about 4 months old and I usually throw one, not more than once a week, from that stage onwards. As soon as the pup's head goes down I walk away smartly, either calling and/or whistling my dog. On its approach I bend or squat down and slightly avert my head, for if you stare any dog intently in the eye you disconcert him. I stroke the pup's head, with one hand under its chin and then remove the dummy. The dummy is always thrown with the command 'fetch it' and in a variety of cover from bare grassland to light scrub or among your shrubs!

To teach the dog to sit I introduce all four commands at once, i.e. long single whistle blast, foot stamp, raised hand, verbal 'Hup'. I walk along with the dog on the lead, stop, give the commands and gently push its hindquarters down. If you keep the lead taut and push with the lead hand downwards then automatically the dog's head comes up. I vary the sequence of command each day. The longest time it has ever taken me to get a dog to sit down voluntarily is three weeks. Initially I only leave the dog sitting for about five seconds, and then give it the signal to move. *It is as important to give the release signal to the dog to move, as it is to give the initial command to sit.*

Once the dog sits freely on the lead I increase *slowly* the sitting time to three or more minutes, and in addition start to drop my lead and *gradually* back away. This is not done immediately but, like the length of the time on the drop, i.e. sit, is done over a period of weeks. I back away to a distance of 100 yards or so and also then practice walking away, looking over my shoulder, ever ready to reiterate my command or give a stern 'no' to prevent movement, until I can walk away out of sight and remain so for about a minute. *Never* at this stage call your dog up to you, for if you do it will soon anticipate your recall signal. Movement will take place, for the pup may think that you are going to leave him. *Never* punish any dog when it comes to you or you will never have your dog returning with a retrieve boldly and unafraid. Gently, if it has the lead on, take it back to the *exact* spot it moved from and reiterate the command so securing obedience. Sooner or later you will have a battle of wills over remaining on the drop, and it is one you must win for it is one of the two keystones of all future training. Every time the dog moves put it back on the exact spot from which it shifted. If the pup has no lead, slip it on and lead it back. If the pup moves when you are close to it, pick it up by the slack

3. The first retrieve.

skin on both sides of the neck and lift it back to the spot from which it moved. Do *not* lose your temper at this stage or any stage of training, for the dog will realise this quicker than you, and once the confidence between you is shattered it will take weeks, if not months to repair.,

You should, once the dog is reasonably reliable in dropping while wearing the lead, take the lead off as soon as the dog sits, and again build up reliability in the exercises suggested above. Do not expect the dog initially to be as good not wearing the lead, as with it on. Once reliability is established try out dropping commands when the dog is running loose at exercise close by. Give only one command at a time, alternating commands.

Once the dog sits on command on the lead, throw the dummy. Up to this time the pup has had no restraint; after a brief wait remove the lead and allow the pup to retrieve, giving the usual command and waving your arm in the direction of the dummy. Continue with this but gradually increase the waiting time between throwing dummy and order to retrieve to about two minutes. Also throw the dummy and walk the dog away, starting from 5 yards up to 2–300 yards, before giving the order to retrieve. Vary also the terrain where the dummy is thrown.

As soon as the dog is proficient in awaiting command to retrieve, questing for 'blind', i.e. unseen dummies, can start. At first drop the dummy unsighted by the dog in light cover and stand about 5 yards directly downwind of retrieve and facing it. Sit the dog which can usually scent the retrieve. Give the command and wave your arm in the direction of the retrieve. If any difficulty is experienced walk with the dog towards the hidden dummy. Build this up gradually to about a 50 yards distance between retrieve and dog. This will be of inestimable value in advanced training and also in the field, with the further advantage that your dog will hunt into the wind on command—all dogs naturally tend to sink the wind.

All these exercises are taken simultaneously and not in isolation, so that the puppy has varied experience and does not get bored. In addition, introduce gunfire and loud noises, preferably at meal times and at some distance from the kennel to start with, so that the puppy overcomes any apprehension it may have concerning this.

I realize I have dealt in depth with the early training but as the late Dick Sharpe and Peter Moxon have so often written, the basic training is the most important and without it advanced and field training are doomed to failure. As an instructor to a dog training club I have come to realize that it is necessary to dot the i's and cross the t's. There is one golden rule always to bear in mind, if you think your dog will ignore you do not give a command, for if you do and the command is disregarded then you have only weakened your control over your dog.

At this stage delivery must be considered, for this can be taught and sloppy delivery in a dog is the result of poor training. If your dog does not carry a dummy firmly or correctly, i.e. at the point of balance, then use a heavy (2–3 lb.) canvas dummy. The weight and smooth surface will make the dog carry the dummy correctly. On delivery do as suggested earlier but as the dog comes immediately to you give the command to sit. Practice in this will result in the dog delivering with head well up, and the habit of jumping up after the delivery of the retrieve will not even begin. Five people spring to mind whose dogs always have perfect delivery styles, they are Mr T. J. Greatorex (of O'Vara spaniel fame), Mr J. Kent (of Chrishall), Messrs J. and K. Chudley (of Harpersbrook) and Dick Male.

The more advanced work can now begin, starting with direction work. Nothing is more annoying than a dog going to retrieve in the opposite direction to the fall. At first use two dummies, sit the dog in front of you and throw them in opposite directions; send to retrieve in order of throwing. Dogs tend to remember only the last, but give a firm arm wave in the direction of the retrieve and, if necessary, move gradually yourself in the same direction. Build this up distance-wise by walking away from the dog up to 50 yards distance before giving the signal. Always walk the dog back to the original spot and yourself go back to your base before giving the second retrieve.

Now throw the dummy and walk away 10 yards from the dog, wave your arm (underhand bowling action) and tell the dog to 'Get out'. Often the dog is puzzled—a new command—so give the known command of 'Fetch it' too. Soon the dog realizes this means to move further away from the handler. Again build up the distance between dog and handler, and dog and retrieve, to 100 yards each. This must be *very gradual* for too much, too often, results in a dog failing to start to hunt until it is some distance from the handler.

Do not allow the dog to retrieve every thrown dummy, but occasionally retrieve the dummy yourself. When the dog is close to you at exercise, give the command to sit and simultaneously throw the dummy—watch your dog! Usually the dog sits; retrieve the dummy yourself and build on this so that the dog will come automatically to sit to a falling object. If it does run in, take the retrieve, sit the dog at the spot where it should have sat, throw the dummy with a firm injunction 'no', from where you originally stood. When your dog does this satisfactorily, you can fire with your pistol at the falling dummy, so getting ready for the real thing.

Throw two dummies in opposite directions and send the dog with a wave of the arm for each in turn, retrieving the first thrown and gradually building up to three and then four dummies thrown to the four points of the compass and retrieved in any order. A thrown ball in rough pasture, through bouncing, will teach a dog to find the 'fall' and cast round for scent and find. Another method is to throw your dummy in the dark; this teaches a dog to listen and mark by sound, as well as to use its nose.

To get a dog to sit anywhere, under any circumstances, is the hall-mark of a trained dog. To obtain this, we already have the dog sitting to deliver, so build on this. Sit the dog 20 yards away and call it up; as soon as it is 7 yards away give a dropping signal. The dog, despite its momentum, should sit about 2–3 yards away. If it doesn't, repeat one or two times; if it still fails, then drag the dog to the spot where it was when you gave the command and repeat the command. Repeat this so that eventually the dog will stop at any spot between its original drop position and you—but beware! This exercise must not be done more than twice weekly, even in its early stages and must be interspersed with walking up to the dog to release it from drop and also recalling the dog to you without stopping on the way in. If you do not do this then your dog will become 'sticky', i.e. will not return to you with the desired alacrity on a recall signal. Another word of warning here, do not expect your dog to return on a recall signal when going away from you; first stop your dog and then after a pause give it the signal. This, too, applies when hunting for an unseen retrieve. The golden rule is first stop the dog and then indicate what you wish it to do.

When your dog sits to your raised hand at a distance, you can introduce dropping to a raised gun. I usually have the dog on a lead and raise a walking-stick gun fashion, stamping my foot at the same time. Quickly the dog associates the raised 'gun' with sitting and will drop to this signal when running at exercise close by.

These more advanced exercises are also carried out simultaneously to prevent boredom and to increase the dog's knowledge, all having practical application in the shooting field. I would, as far as possible, repeat this work in as many different spots as possible for I have found that, unlike spaniels, Labradors require as wide a variety of experience in different surroundings and terrain as possible.

Jumping and water work can be carried out as soon as one starts the basic training. First get the dog to jump over small objects 12 in. to 15 in. high—an easy way to do this is to make slots for solid boards to be dropped into, on the door posts of the run, so that to get out for exercise the dog has to jump. Wedge the boards to make a firm jump, for dogs tend to jump on to the object and then jump off it. On negotiating the jump give the command 'over' so the dog comes to associate jumping with that command. The jump can then be built up to 3–4 ft. as desired.

Once jumping freely then the element of retrieving over an obstacle can be introduced, first using a light and then a heavy dummy. The latter is necessary, for when carrying a heavy retrieve all dogs, unless taught, tend to hunt for a hole in the fence. Whenever the opportunity arises practice your dog jumping pig, sheep and poultry wire—a dog can see through it but cannot get through it! Do not encourage the pup to tackle barbed wire, for a mistake in jumping this results in a gash at the very least. Stone or brick walls of a suitable height should also be included in the jumping curriculum.

4. Jumping freely over boarded jump.

Despite the Labrador's origin some have to be taught to swim! The best method is to introduce the dog to a brook or shallow stream where it can paddle on a firm base. After this choose a pond or brook with a gently shelving bottom and throw in a dummy preferably not so far that the dog is forced to swim. When confidence is assured then throw in a dummy that requires a few swimming strokes. If this is satisfactory, leave well alone and repeat later gradually increasing the exercise so that the dog will swim out for two or more marked and then unmarked retrieves. What must be practised is the hunting for unseen retrieves beyond the far bank of a river. Practice with marked retrieves first and then blind retrieves using the wind. This, plus hand-signals when the dog will accept these, should enable you to recover your game when it has fallen in these circumstances. If a dog won't swim or go more than belly-deep in water, the example of a free-swimming dog may help, especially if you allow both to go together for the same retrieve (jealousy). If this doesn't work then go to a canal, put your dog on a lead then lower it into the water and gently tow. Half-a-dozen repeats of this visit does the trick and instills water-confidence in the dog.

The three final parts of this advanced training are handling at a distance on three dummies, walking to heel and stopping on going for a retrieve. I teach the first by sitting the dog on the base line of a triangle, the dummies being the three vertices. By hand signal (I am beyond the dog, only a few yards distant) I get the dog to pick them up in any order. I also do this with hidden ones, building up from two hidden in the opposite direction to all three.

To stop a dog by whistle on its way out to retrieve is easy, if the other work has been properly done. Merely throw the dummy and give the command to retrieve, but when only five yards on its way, command and whistle 'Hup'. The dog sits—if not repeat the correction after delivery, of taking the dog to the spot where it should have sat, giving the ignored command. Gradually build this up so that the dog will stop at any distance

5. Water; the finished product.

between you and the retrieve. Again be sparing in this lesson so that the edge of retrieving is not blunted and intersperse this lesson with all the other work.

To walk to heel is taught in a week or so. Labradors are natural 'heelers'. When walking on a lead jerk hard so that the dog's head is by your knee and utter the command 'heel'. As soon as the dog walks to heel, let the lead trail or wind it collar-wise round the dog's neck. Build up from 5 to 200 yards walking like this and then do the same without the lead. If the dog walks in front, gently pull it back by the scruff of the neck, giving a command, and the same if it lags. Build up so that, except at exercise or working, it knows its rightful place is at heel. Leave this until the last because too much heeling when the dog is very young inhibits the pup and slows its drive.

During this advanced training period I have done two other things; firstly, introduced my dog to ground game and also to retrieving dead cold and warm retrieves. Your puppy must be chase-proof to ground game. If I have no access to a rabbit-pen (a wire enclosure in which rabbits live natural lives) and I let my dog on a lead sniff a tame rabbit on the lawn hopping about much can be achieved. Once the dog accepts the rabbit, sit the dog and drive the rabbit near to the dog. When I am happy that the dog is reliable I throw a dummy near the rabbit for a retrieve—you must ensure before this stage that your dog will ignore the bunny! This takes place over several lessons. Once your dog is reliable here, it does not mean that it will behave in the shooting field. My own puppy brought up on these lines, was the only dog (and a no-slip retriever) at three recent hare shoots during which some 400 hares were shot. She proved as steady as a rock, but I did *not* let her retrieve a hare! It is useful, too, if you get to know a pigeon fancier who will allow you to take your dog to watch his birds flying in and out of the loft. Better still if he will feed them near your dog so that it can watch. Always watch your dog and keep it on a lead until you are reasonably sure it will not chase.

The first game retrieves depend upon the season, but I try to give retrieves of all the game the dog is likely to meet, always dead, first cold and later warm. With pigeon, however, I first put the bird in a nylon stocking, so that my pup can make its first retrieve or so without the loose feathers associated with this bird. I either throw it as for a normal retrieve or drop the retrieve, walk on and send the dog back. As soon as my pup's head goes down I walk quietly away giving the recall signal. In 99 per cent of my dogs all goes well. If the dog appears reluctant to deliver I go out of sight—no dog likes being left behind. If it won't pick up the retrieve tantalise the dog with it, playing with it and getting the dog roused and excited. I then throw the retrieve and allow the pup to run in.

The first few trips to the shooting field can, unless you are careful, undo all your training. I always keep my dog on a lead and watch him carefully, for this is a new environment. Care taken now will pay dividends for the rest of his life. Do not keep him out all day—for the first few times I only have my dog out for half a day—and again on his first trip or so, do not let him retrieve. Even after that only allow him to have selected retrieves and never of those birds that are in full sight and easily picked up. After all, a dog is for those retrieves that you cannot yourself make. As a finishing school for steadiness I have found pigeon shooting invaluable. You have the falling birds, the shots, pigeon overhead and you sometimes leave your dog while you stalk a bird. On these occasions I do not allow my dog more than one retrieve and that usually of a dead bird.

Do not expect him to be fool proof; he lacks experience and if he is properly trained then his mistakes will be solely due to this. Never send your dog for a runner in view, and only then for those birds heavily wounded. Lastly, if in any doubt as to what to do sit your dog and think; when hunting for a retrieve trust your dog for if you could find it you wouldn't need a dog!

BREEDING

After a few seasons you may decide to breed from your Labrador, dog or bitch. To obtain the best chances of success you must line breed. If you have a dog you are really dependent on someone bringing a suitable bitch to your dog. However, with a bitch you can take her to any dog you choose. If you originally chose with care, the choice of a mate is relatively easy, remembering that *like breeds like.* All matings are either random breeding (two unrelated animals mated), in-breeding (mating of closely related animals e.g. father — daughter), line-breeding (modification of inbreeding) or outcrossing (mating of unrelated dogs, one of which is closely line-bred or inbred).

To produce what you require, the best bet is to line-breed to one or more outstanding dogs in your bitch's pedigree, i.e. have these dogs in the sire's pedigree too. To random mate is like 'winning the pools' and the chances of producing what you want are thousands to one against. To inbreed, whilst it has some advantages, you need for success to have a great knowledge of both sire and dam and their 'family' history. Not only in this mating will the good points come out but so also will any faults be accentuated, and drastic culling may be necessary, for the first generation of this mating may bring to light unsuspected qualities or faults. Outcrossing, i.e. an inbred animal mated to an unrelated one, may be good kennel-wise but to buy from such a mating is perhaps unwise, for the outcome of hybrid vigour is evident in one generation only.

After mating takes place, usually around the 12th day, ensure that the bitch is kept safely until her period is finished. Labradors usually give little trouble in whelping and in these days of infra-red lamps it doesn't matter when the pups are born. Allow the bitch time to get used both to lamp and whelping quarters, which must have a wooden floor or base. Give her plenty of newspaper—to be torn up when she makes her bed. Remove this after whelping, substituting paper sacks, clean hessian sacks or a blanket. New born pups are best left alone, after a cursory examination to see if any are malformed. If she has more pups than teats, then you must consider reducing the size of the litter.

Feed the bitch at first with milk, enriched by glucose and calcium, and keep this up throughout her nursing time. Raw meat is a must for this promotes a good supply of milk. At 3–4 days have the dew claws removed by your veterinary surgeon. From about three weeks onwards leave the chosen meal and plenty of water for the pups to start to feed themselves *ad lib* and introduce them to raw meat, by giving each pup, each evening, a little minced meat. In the morning I give them milk plus cod-liver oil. From five weeks onwards I introduce my puppies to tripe and find that that they enjoy pulling and tugging at it. I give them milk twice a day. If tripe is unobtainable, then feed raw meat once a day, giving milk also once daily.

I remove the lamp at about five weeks old unless the weather is severe, having hardened the pups off from three weeks onwards by gradually reducing the time it is on, reserving it eventually for the nights only. From 5–7 weeks the puppies are gradually weaned by at first leaving the bitch with them at night only, and finally allowing them to suckle for not more than a few minutes daily. During the period 3–8 weeks old is the *vital* period of 'humanising' when the time *must* be spent daily playing and handling the puppies (children are an invaluable aid in this). At six weeks old they must be wormed for round worm irrespective of whether they show signs of worms or not; do this after consultation with your vet. I need hardly say that during this time your bitch must be very well fed.

I now use paper sacks for bedding and find that as the pups grow they foul only one area; this is easily cleaned by removing the dirty sack and substituting a fresh one. At eight weeks old we can commence the cycle again towards the production of a trained companion in the shooting field.

2 Golden Retriever

Like all gundogs the Golden Retriever was evolved in the hands of breeders whose sole interest was in their working capabilities. That they were also men of means, whose aims were quite uncommercial, resulted in the foundation of a breed excelling in its field of work.

The Golden Retriever, as we know him today, is enjoying ever-increasing popularity as pet and show-dog. The price of this popularity is a decline in his working ability—the penalty of the ready market available for puppies and the deplorable tendency to breed from immature and unproven stock.

Although registration of the breed 'Golden Retriever'. or Retriever (Golden) as it then was, was begun just before the first world war, many pedigrees can be traced from the present day back, at least, to the beginning of the century. At this time they were known as wavy-coated retrievers and only distinguished by colour, but there were many registered dogs so recorded. For practical purposes, therefore, pedigrees are traceable so far, but beyond that the researcher goes into unchartered country and is tempted to conclusions which are not backed by evidence.

The well-known story of the troupe of Russian circus dogs bought in Brighton by Sir Dudley Marjoribanks (later Lord Tweedmouth) and sent *en masse* to Guisachan to found the kennel, has gained credence more by its romantic appeal than for any other reason. Lord Tweedmouth kept stud books and records of all his dogs for a period well-covering the date usually accredited to the circus dog episode, and no mention of these animals can be found. They would, one feels, have merited an entry in those well kept records. Many breeders, notably the late Mrs Charlesworth, were enthusiastic champions of this theory and her books were unequivocal on the subject. However, there were always some who were not satisfied, and, albeit reluctantly, the circus dog story must surely give way to the patient research and conclusions reached by Mrs Elma Stonex, conclusions backed by sound evidence and with access to the Guisachan papers.

It is well known that there was a breed of yellow or liver coloured dogs of spaniel type in the Border country and notably along the Tweed. These dogs lived aboard the fishing boats and were well known along the east coast at the ports where fish was landed. It seems clear that the Newfoundland dog, who led a similar life there, on the Cod banks, and was enthusiastically described by 19th century travellers, was brought over to improve the border breed. One feels that 'yellow' or 'liver' could fairly describe almost any colour of coat subjected to constant salt water. I have spoken to many folk in the 1920s and 30s who remember the 'ship's dogs' and used to regale me with tales of how they saved lives by swimming ashore with a line when the boats were in trouble on the rocky coast.

Lord Tweedmouth is known to have founded a breed of yellow retrievers using two dogs from the Tweedside area and described as 'Tweed Water Spaniels'. One of these, Belle, mated to a yellow retriever belonging to Lord Chichester, produced the four yellow puppies which authorities on the breed consider to be the true foundation of the Golden. This litter was born in 1868. One of the bitch pups went to Lord Ilchester and began the kennel there, another foundation stone of the breed today. Until a very few years before his death in 1894, Lord Tweedmouth bred with a deliberate policy of establishing a strain of yellow retrievers. These dogs were bred at Guisachan, in the Highlands, and over the years there, and at neighbouring estates, they continued to breed from the pups which came from the original Tweed Spaniel/yellow retriever foundation stock. In the last century and in this until the sprays, silaging and tractors of modern farming appeared, the grey partridge abounded in most parts of Britain. Shooting over dogs and later, with the coming of the breech-loading gun, driving the birds, probably constituted the major part of game-bird shooting then practised. Until the last war there was at least one covey and usually more in every field in my area; now, alas, partridges have virtually disappeared from the district and from many others. Setters and pointers were used to locate the coveys and for this purpose 'nose' was of paramount importance. It was, therefore, natural to find breeders turning to these dogs when an outcross to improve 'nose' seemed indicated in retrieving breeds. At least one setter appears in early Guisachan pedigrees.

In those days performance was the sole criterion of worth; present day obsessions with 'type', and breeders with one eye on the pet/show market for economic reasons, simply did not exist.

The late Sir Watkin Williams Wynn had in his kennels at the beginning of the last war, a few dogs left of a breed he had evolved by crossing wavy-coated retrievers with red setters. They looked very like goldens and I saw two of them working many times. While few of these dogs left Wynnstay they have occurred in pedigrees and must be given a place in the foundation of the breed.

My first golden had Guisachan blood in her pedigree and while this bitch showed no tendency to 'set', several of her descendants did set momentarily before picking a bird. In the last 20 years only one of our goldens has given a hint of this echo of a far-off ancestor but I have no doubt many instances have been noticed by handlers.

As all animal breeders are aware, one is occasionally startled by what is commonly called a throwback, when some feature of appearance or behaviour seems utterly at variance with immediate ancestry. Similarly two litters of dogs, identically bred from a repeat mating, will occasionally produce a variation of looks and ability. A timely reminder that nature still keeps a few secrets. It also emphasises the importance to a breed of only perpetuating the species with dogs which truly measure up. In these days of hard commercial thinking and costly feed bills, quick profits override integrity and progress in breeding is undermined and greatly hampered.

Golden Retrievers are relatively slow maturing dogs and most are three years old before they are fully developed. People will not wait, perhaps in fairness cannot wait, nowadays which makes one realise how much we owe these pioneers of the past whose careful selections and breeding programmes knew none of our modern pressures. Economically, too, their task was simpler, their aim more single-minded—to produce the best game-finding animal to accompany them on shooting days, and to track and retrieve wounded game tenderly.

BREEDING GOLDENS

To breed a litter is, to some people, just something a female golden will do if mated to a male, with the vague hope of some monetary gain to the owner in a few months' time. If I may assume that anyone reading this has some interest in the breed, there are points which are worth remembering. If you have bought a bitch pup originally from a reputable breeder of working stock, if the bitch has proved she is a good worker, has the charm of her breed, devotion to her 'family' and has reached maturity with ever increasing pleasure to you—then breed from her. It will be found that some of the benevolence of the golden rubs off on their owners, and the breeder of your bitch or someone who has some years of experience behind them will whole-heartedly enter into the sire selection exercise.

Truly working and worked goldens are so very thin on the ground compared to the show dogs that the choice is narrowed. Against this those that have working dogs will know well where a suitable sire can be found. A heat early in the year is the obvious choice for a shooting dog; the pups will be old enough to start work themselves after another season and the bitch fully recovered and fit by the time you need her again. Young golden bitches will sometimes only stand to a dog for a very few days so it is necessary to watch carefully for the first signs of oestrus. The eleventh to fourteenth days are, I see from records, the most usual for a successful mating the first time. (The all-time long, I see, was a marriage conducted on the 21st day of heat, from which a bitch's fifth litter resulted.) One mating is sufficient if a tie results and the average time for

6. A busy pair of working Golden Retrievers.
G. L. Carlisle

this is 18 to 23 minutes. If, however, a second mating is wanted I believe that 36 hours is as long an interim period as is desirable.

Pregnancy can sometimes be confirmed visually three weeks later, but unless one is experienced it will probably be only after six weeks that the signs are obvious in a first time mother-to-be. Health during this time is usually very good but check any temptation to over-feed. The one meal a day routine should be maintained, with added calcium and vitamins, but the food can be divided and fed twice if the bitch gets very heavy towards the end. Proprietary full-cream milk foods are excellent and it is a good plan to get the bitch taking this happily before the pups are born.

Most important is the choice of a labour ward. Even the most placid family pet will often exhibit signs of wanting to go off and dig out a rabbit hole or get into some inaccessible dark corner when the pups are imminent. The instinct of any mother animal whose young are born blind and are helpless for the first few weeks of their lives is to hide them up. In goldens this instinct can be overriding—overriding the devotion to and apparent complete dependence on their owner.

The place chosen should be a quiet loose box or shed well away from other animals. A four foot square bench with three high sides and one low one, sited where a heater/lamp can be suspended above it, will meet the family's needs for the birth and until they are weaned. The lamp should be hung by an adjustable chain so that it can be raised and lowered quickly and easily. If the bitch is a house pet it is kinder to start in very good time to let her become familiar with her quarters. Short periods by day should eventually become the whole night when she has accepted the change. It is as well to have the heater/lamp on several times as the warmth is a factor in getting her to adopt the bench.

Usually the first litter is born on the 61st or 62nd day after mating; subsequent litters are born probably one or two days later than the first, but this pattern can vary with individuals. Refusal of food is often the first sign and newspaper on the bench will be torn up and arranged by the bitch as labour starts, thus satisfying the instinct to make a bed. I always stay with a bitch when she starts whelping and see her well on the way. Should all go well, and it usually does, she will attend to each pup as it arrives, frantically licking to clear the membranes and gnawing through the cord. With goldens, which normally have fairly large litters, the lamp is invaluable for keeping the first arrivals warm while the mother is busy dealing with the latest birth. The remains of the newspaper can be quietly removed at this stage and the bare, not too smooth, boards let the pups crawl up to the bitch.

When she has finished and cleaned up, a very small drink of milk food should be offered and then she should be left in peace. Pups should not be handled at this stage; in a few hours she will come off the bench and go out when a quick survey can be made and sexes counted. For at least 24 hours only liquid food should be given, then gradually increase her food until she is having her main meal at the normal time again, supplemented by Lactol first thing in the morning and last thing at night.

It is very important that the bitch should not be worried by strange voices or other dogs coming near her, and no one but her owner should visit her for the first four or five days at least. Bitches will often lie perfectly quiet and apparently unperturbed while admirers coo over the litter, but after they have gone, in some cases long after they have gone, the old instinct takes over and the bitch will seize a pup and begin a nervous search for a new home for her family. This is when there could be a disaster. Being shut in she will have to settle for somewhere nearby and, probably, quite unsuitable. This worries her. She will start moving them one by one but at this stage she is acutely sensitive to the pups' squeals and the first ones deposited in a new cold home will protest very loudly. In a very short time she will be rushing back and forth, now with a split family and, in a frenzy of motherhood licking the yelling pups. Should this situation get out of hand she may re-double her licking and finally eat the pup. Once this has happened the casualty list mounts rapidly and all your hopes may, quite literally, disappear.

After the puppies are dry and settled the less noise they make the happier you should feel. If all is well the family should be virtually silent apart from a few contented mutters. Prolonged squealing from any or all is indicative of trouble and should it continue call your vet at once.

Should the bitch have any difficulty in passing water you will need professional help, for this can well indicate an unborn pup restricting the passage. An injection can restart labour and the pup be produced without recourse to surgery in some cases, but the condition must not be neglected. Therefore it is wise to watch for the resumption of normal functions following the birth.

If you have more puppies than is advisable from the bitch's point of view or your own, cull at the third or fourth day. One of the least agreeable facets of breeding, postponement of this chore is both cruel and unnecessary. Make up your mind and slip the surplus pups out of the nest while the bitch is out or diverted and have them humanely destroyed without delay. Clip off any dew claws on hind legs only; the dew claws on the front legs should never be removed. Watching carefully will show how these claws are used when a dog is climbing out of a steeply banked river or pond or up a steep ghyll-side. We also see them helping when the dogs are negotiating a high wall, the last foot or two of which has to be climbed. Many where I live are between five and six feet high.

The growth of puppies is very rapid and the bitch should have her food increased as required to as much as three meals a day plus the milk food in the morning and last thing at night. In ten days the first glint will be noticed and 48 hours later the eyes will be open and the pups become more mobile daily.

If the situation calls for it the family can start lapping at three weeks, but if the bitch is doing well another week can pass before starting them off. Make up the milk food—I have used Lactol for many years—according to the instructions on the tin and, at blood heat, put a little in a shallow dish, dip the pup's mouth gently into it when his reflex action of lapping will start at once. He will draw back spluttering and get the taste of the milk; another dipping session will get him going, and the process is helped by the pups licking the milk off each others faces. The next step is to start the family on Farex, Saval No. 1 and minced meat. At all times the food should be fed at blood heat (101°F.). I am convinced that this stage of rearing is of paramount importance to healthy youth and a fit young dog. A well reared puppy gets off to a good start and as his destiny is work well-laid foundations pay off throughout its life.

Golden Retriever litters are very active; if the management is good they are either in a heap asleep or investigating their surroundings with ever-increasing curiosity and boldness. Over-done, overfed puppies are lethargic and too heavy for their legs, they seem to lose the sparkle and quick reactions so essential when training starts in a few months time. At five weeks worm the litter; with modern drugs this essential operation is easy for you and for them and should be repeated two weeks later.

Weaning the litter is a gradual process and bitches vary in their devotion to the family. I leave the way clear for the bitch to come and go as

7. A Golden Retriever carrying a partridge takes a gate in splendid style.
Scottish Daily Express

she likes at three weeks. By four weeks she will probably only spend the nights with them; by six she will be disinclined even to do this and by eight all suckling will be over although she may return to regurgitate food to them but will probably dodge away if they try the milk bar. From first lap to complete independence of their mother the puppies should be fed little and often. Four meals a day and a drink last thing at night will be the routine by the time they are ready to go to their new homes.

This is a major step in the dog's life and can be accomplished perfectly smoothly. Give each purchaser a diet sheet of what the pup is having but impress on him that this is only for the time being. I was horrified to be rung up by a worried owner and found that she had faithfully adhered to the breeder's instructions for an eight week old until the wretched pup was over five months. A good tip for the first few weeks is a small tea chest into which the pup can be put at night. Draught free and fairly dark he will be much more inclined to settle than in a larger area where he can wander off. They seem to miss the other pups less and find comfort in the tea chest for the first period of separation.

By four months there should be only three feeds a day and a large marrow bone makes a good toy and an aid to teething. At six months two meals a day and any time after ten months the one meal a day, which will be his routine thereafter, can be gradually introduced.

In addition to the diet sheet each purchaser should be given a signed pedigree and transfer form, registration card and a plea that the pup will

8. A Golden Retriever, alert and ready for action.
John Tarlton.

have his anti-distemper/hard pad, etc. injections, as soon as the vet will do it. Protection is vital. Distemper exacted a heavy toll in my youth, and I suppose that 50 per cent or less of dogs were reared. Almost without exception every litter got distemper in those days when good nursing was the only known aid. Provided complications such as chorea, paralysis or pneumonia did not occur you could usually get them through. If the disease was in a mild mood that year a high percentage survived, but in a bad year so many died and so many developed the dreaded damage to the nervous system that losses could be very high. Always you were asked 'has he had distemper?' when selling a pup.

Now, because the protection is so good people forget, or possibly do not know that distemper is endemic and, particularly in towns in the months following the Christmas present puppy buying, frequently assumes epidemic proportions. In these days of motor car mobility it is so easy to run into infection that it is prudent to keep your youngster from meeting other dogs until immunisation is effective.

TRAINING

When early man domesticated the wild dogs which probably followed nomadic tribes for the bones and refuse, he began a partnership which has truly endured. Training a dog for use with modern shooting methods is adapting a natural ability. It should always be borne in mind that a dog is a pack animal and that his instinct to obey and follow the leader of the pack is the instinct which makes him look up to and follow his human 'leader'.

In field observations of wild dogs scientists have seen that 'pecking order' or precedence is a built-in factor of survival. Our dogs do not learn their craft or status from canine tutors but from us. It is largely the psychological understanding of the dogs' instincts which makes the top class handler and causes those men and women who have this flair to succeed, year after year, in competition at the expense of their less gifted rivals.

A puppy who can be given maximum freedom in the first months of his life will learn much to his advantage of environmental knowledge. He is adapting to human influences and a human dominated world. Once training has started for the particular role he has to play in our sport as presently conducted, he should be shut up. A young puppy has very limited powers of concentration and when his lessons start he will be much more able to learn, mentally and physically, if tuition co-incides with the short times of being taken out by his 'leader'. It is futile to expect a ready response to any new step in education from a pup who is already tired from romping on his own or with human or canine friends.

Retrieving breeds should never exercise their love of carrying with children, and under no circumstances should anything they pick up be taken from them roughly, nor must they be chased to give up something they have decided to play with. Early errors of this sort make real problems for the trainer and can even destroy any hope of a final gundog. It is true to say if he will carry he will retrieve, but it is the skilful handling of this instinct which makes the final product.

A retriever having made contact and picked up a bird must have one object in view and one only, to deliver it into the hands of his 'leader'. It may well be that a wounded duck or a running bird will take the dog many, many yards from where he was sent, and once he has achieved the find, only one thought should be in his mind. Good training will make sure that he comes as straight back as circumstances permit. In fact, his eagerness to return is a measure of the confidence he feels, and stresses the necessity of all completed and praised retrieves being the dog's most cherished achievement. Hence *never*, *never* should any early carrying incident be associated with a reprimand.

If I had a £1 for every time I have been asked 'when should I start teaching my pup this, that or the next thing?', I should be very well off! Goldens, like any other breed, vary a great deal in their rate of development. One of the earliest successes in terms of age that I can remember was a field trial win by the Golden bitch Dragon of Tyrconnel in the mid-twenties during October of the year she was born in. As she was whelped in April this was a remarkable achievement and, from memory, I think the record stood until quite recently. One of ours ran second, in a 16 dog stake three weeks short of his November birthday but this is not an uncommon age to be in the frame though far more the exception than the rule.

Professor Konrad Lorenz writes that a dog selects his 'leader' from about five months onwards and a puppy must be ready in this sense before training can really start.

The basic requirements of hand breaking are walking at heel, sitting and staying on command, retrieving to hand when bidden to do so, and responding immediately to the whistle or voice. Goldens are cheerful characters and imbued with a wish to please to a very noticeable degree. Time spent on the early stages of training is very well spent; sometimes patience in waiting until the right moment to start a new lesson will account for much of this time, but a puppy whose progress is smooth will progress faster than one who has been asked to run before he can walk. Undoing a bad habit, sometimes so quickly learned that you are taken unawares, is infinitely harder than teaching a new step in education.

Introduction to swimming is often just a matter of visiting a pond or river, provided that the initial dip is not, in any way, a frightening experience. All Goldens, of course, can swim and most love it. Shallow water and a free-for-all play will often get the pup initiated in the most natural way and it is seldom that this new element needs more than a commonsense approach. Some years ago a pup fell off a high bank into a deep pool in the river. Surfacing, he swam spluttering to the nearest point and showed a strong distrust of water for many weeks afterwards. When finally paddling, then wading with him, got him going again, he suddenly took to it and would go off and solemnly cruise about as though to convince himself that he could cope. He was no further trouble but the carelessness of allowing that first fright cost a lot of unnecessary time. A bold pup, he completely threw off the effects, but a more timorous one could have been unreliable in water thereafter.

A leading trainer told me the following story against himself. He was training a springer whose enthusiasm for water was distinctly lukewarm. He shot a mallard into the middle of a low river and, rightly expecting the dog to get it, was very put out when the brink was as far as he would go. Cajoling and ordering failed so he put down his gun and threw the dog in. The mallard was very slowly drifting away. The dog swam angrily to the far bank deliberately passing the duck, where he sat and gazed at his handler; nothing would make him move. Eventually the trainer returned to his van and made the four mile trip by the nearest bridge to retrieve the dog, and the duck was lost. The point of the tale being that all aspects of training must be complete before you have a gundog, and that it never pays to lose your patience.

Living, as we do, where rivers tend to be fast and often in spate, we try to get the pups going confidently into broken water in the summer before the shooting season starts. A dog should, ideally, come out where he goes in but he must also learn to use the current and make a landing more easily and quickly than he can by fighting against the pull of the river. I saw a good dog nearly drowned by her obsession to swim back up stream with a bird to where her owner was standing. She finally abandoned the bird in a panic of yelping and was rescued by a quick-thinking keeper who had run for a boat tied up some distance upstream. She was used to still ponds and slow rivers and this, to her, new element nearly cost her dearly.

After a season our dogs learn to get ashore as best they can, often a long way downstream from the entry, but a little practice is helpful and few seem to go with the current instinctively.

Today transport is another aspect of training. Most dogs take to car travel happily and are always eager to go in a vehicle. The odd one is car-sick and this is a nearly hopeless condition for a gundog. The worst one I ever had was already slavering before we left the yard and even the nearest meets were a nightmare. Someone suggested a method of overcoming the trouble, so simple that it seemed worth a try. We took the doors off the Land-Rover and fed him in it, letting him jump in and out at will. So far as I remember this went on for about six weeks. He was completely cured and lived out a useful life of 14 years. I passed the tip on to friends with a charming young Labrador similarly afflicted; they used the boot of their car and this bitch became a happy traveller. If this does not work I believe modern drugs are safe and effective, but cannot speak from experience.

In stock country most fields are fenced with wire. Pig netting topped with two strands of barbed wire, or two strands of barbed wire above several plain wires are commonplace. We teach our dogs to jump from an early age. Some are naturals and do not often 'meet it wrong' but if they are sent over fences regularly the chances of damage are lessened. Barbed wire can inflict very painful injuries and cause the dog to be off work for weeks. A small-mesh netting fence is a good starter and the pups soon learn to jump it cleanly. If you watch carefully you will see the manoeuvre completed with ever increasing skill and ease. Many years ago I was sent in a field trial for a 'dead' pheasant which, shot earlier, had carried on for several fields. When the line arrived within 50 yards of the mark I was told just where to send the dog over a fence on our left into a stubble field. The dog found the fall and was off on a line very quickly. We saw him next 200 yards in front of us going on a strong scent. Soon he was away on our right from where he re-appeared with the runner. Before we could do anything we suddenly saw immediately in front of the dog another fence of five strands—and every strand was barbed. Of all dogs, had I had the choice, this was the one I should have chosen for this situation which, under the circumstances, was no one's fault. He was a super jumper and gave that ugly fence a foot. Had any dog tried to slip through that fence it would have been terribly injured. For many years after that judge never failed to remind me of this episode when we met.

Some dogs will 'bank' over barbed wire and get very clever at missing the spikes. Should a dog get hung up by dropping a leg in a slack fence get to him as soon as possible, push a stick in front of him to bite in his pain and panic, and try to lift him over the fence to the take-off side, when he will usually free himself, as you support his weight. Some years ago my dog got hung in a trial coming back out of roots down a hill. I got to him as soon as possible but, of course one's hands are empty at a trial so it was lucky he made no attempt to bite although he was roaring with pain and fear. Help was at hand and four strong men took over but it was a good three minutes before he was freed, and only after a post was wrenched out. He was not apparently injured but next day and for some weeks he was unable to jump into a car as he had strained his loins. This Society deputes a member to carry wire-cutters at all trials since this episode, and it is a sound precaution. No single person could have freed this dog; the fence was slack and in front of a deep ditch so we worked with some 60 lb. of dog held above us until someone could climb over to the other side and take his weight.

On a shooting day your dog is more than likely to be hustled into a vehicle with precious little room for canine and human passengers. Temperament is invaluable here and, provided the first few experiences are pleasant there will be no trouble. Once two overweight black retrievers became locked in mortal combat in a very crowded van on a grouse moor. It was difficult to keep a straight face in the following vehicle as elderly and young guns alike cascaded out of the *mélée* on to the heather. Picking themselves up the owners of the combatants angrily accused each other of having a fighting dog. It looked for a moment as though there might be a human reproduction of the battle still raging in the van. When the dogs were separated and order restored there was a re-arrangement of passengers and the convoy proceeded. Only over lunch was complete rapprochement achieved—it is difficult to be bad-tempered on a good shooting day.

Golden retrievers are not guard dogs and any shooting dog with a tendency to guard property or game is a potential menace. Young male dogs sometimes go through a very bumptious stage and become self-assertive and inclined to raised hackles for a brief period of their development. Judicious checking of any signs of belligerence will usually end this and it seems to be a passing thing, noticeable between about one and one and a half years of age.

We have never had to discard a Golden for ill-nature to other dogs or to people. Good nature is an integral part of golden character and an essential in his temperament.

PICKING UP

For me, like so many other retriever owners, picking-up is the most rewarding and enjoyable thing I do with the dogs. Any young handler who cares to take the trouble to train a Golden carefully can have endless fun and interest and do a useful job. It is very important to get your dogs really fit in July. As a picker-up on a moor you are frequently called on when a gun has searched the area and is one or several birds short on his count. Even if your dogs have spent time away behind the line on long and towered birds, they should still be fit enough to take over when the guns' dogs are stretched out panting and clearly unwilling to continue the hunt. I am sure good feet are a valuable asset here, the round cat-footed dog will continue with ease where the less well equipped animal will, especially in areas of burnt heather, soon be reluctantly picking his way.

Scent on a moor, when it is at its most beautiful with the heather in full bloom, can be nearly non-existent. This condition prevails more persistently if the weather is hot in August and early September. As the dogs hunt, a little white plume of pollen rises about them and the best nose in the world is virtually put out of action. One can speculate that this is nature's way of protecting against predators, for a gundog, like foxhounds on a poor scenting day, soon realises conditions are against him and enthusiasm wanes. A predator would not choose this hour and these conditions well knowing his prey was too hard to find, and the prey would have its respite.

A well trained dog is also an enthusiast and it is not hard to get the response you want; the extra bit of effort is forthcoming at your behest, when the man-dominated situation seems, no doubt, pretty hopeless to the dog. The success saga which you have stressed so much in training really pays off now.

A good picker-up should add at least 10 per cent to the day's bag on a driving day. Your host will brief you on return drives and where you may stand, ideally out of sight, for grouse see far and you can well cause birds to deflect and fail to cross the butts if you are wrongly placed. Should the terrain permit you can be quite close to the line, say behind and below the butts in a gully. Your job is to watch for any wounded birds which come through and fall a long way behind. Make sure you have a good view of a lot of moor. Practice improves marking and the line is the important point, rather than the distance. Get a line on your birds using the distant outcrop or a green patch beyond the fall—not on those sheep so conveniently placed at the time—and note each one down. When the beat comes in have your plans ready and get your birds systematically. A dog should handle straight out for up to 150 yards and your and his energies can be considerably conserved and much time saved if you have a good fix on the area he must go to and the side he must be on to use the wind. Should birds have dropped near you the dogs will have been able to mark them and you will be able to use the dogs together, indicating to each which bird you want. Picked up behind, you can then go to the butts where, even if everyone is satisfied, it is always worth hunting down the line if time permits.

Many a bird is overlooked near the butts through poor counting or neighbours' dogs trespassing. Don't blame the gun for poor marking and if he is shooting two guns and has a loader don't blame him either! He has a job which takes up all his attention and neither man can afford the time to watch all birds down. If you are told there are missing birds in front of the butts it's always worth just not noticing when the dog works his way behind—it's more than likely the birds are there.

The older I get the more convinced I become that we far from understand all our dogs' equipment. We have a dog now, alas 12 years old, which has picked countless grouse. Waiting behind with him he, like me, watches the coveys come through and if all appears well with them he, and I stop looking. Time after time he has gone on watching for as long as they are in sight and time after time when he has done this a bird has peeled out and fallen or towered in the far distance. It happened so often that I rather diffidently mentioned it to a keeper of some 70 summers with a lifetime on grouse moors. He was immediately interested and told me that he had had at least three dogs over the years, which, like ours, knew when a wounded bird was amongst the others although, to our eyes, it could not be singled out by its flight.

Since this conversation I have been emboldened to discuss this with others and have met several people whose dogs do this with pheasants and concentrate on certain birds which, apparently unscathed, glide into a covert and will be found to be dead if you go for them. I have a son, two grand-daughters and a grand-son of this dog's, now working on the moors but cannot say I have seen any of them with this instinct. What sense is used? It cannot be sight or we would see it, it can hardly be scent so far above and often down wind. Sound can also, I feel, be ruled out as there may be several shots at the relevant moment and rattling pellets don't necessarily induce it. I would love to know but can offer no ready explanation nor could my keeper friend.

At all times when a picker-up remember the Victorian child—seen and not heard. It is quite unforgivable to be rushing about waving arms and blowing whistles. Only fully trained retrievers should be used for this job. A young dog who is sufficiently reliable can be taken out with older colleagues and, with discretion be given a stone-dead bird or two, but you do not train your dogs at your host's expense.

A horrid example comes to mind. Two ladies camping nearby in summer, persuaded a keeper to let them pick up in the season. They arrived half-way through the first drive. There are eight stands at this shoot and the bag is in excess of 400 pheasants. Never, said the dogs, have we seen anything like this and they yelled their approval as their owners were nearly airborne. The drive over the two dogs were released. They streaked off to the nearest gun and joyfully picked up bird after bird abandoning each one in favour of the next. The gun, who had undone his treasure's tether, stood open-mouthed and in obvious danger of a seizure. As a guest he presumably felt unable to speak his mind, which, at that moment was probably just as well. The keeper was interviewed by our host; a kindly man, he found me and my fellow regular pickers-up and consigned the nightmare to us for the rest of the day. We can't have done too badly, for at the end two beaming ladies accepted a brace apiece and departed. The story, losing nothing in transit, went the rounds and as one of the dogs was a Golden, my leg was unmercifully pulled.

It is very necessary to know the ground to be efficient at a big pheasant shoot—the coverts you may or may not go into and where to stand out of any gun's way and so positioned that a runner can be cut off before it makes its way into forbidden territory. Your job is to collect birds that are (a) wounded and run, (b) carry on and fall far from the line. You do not collect dead birds lying near a gun who probably wishes to pick them himself, or if he does not the keeper and his staff are responsible for their collection.

It can happen that a running bird makes its way into a cover already shot or not on that day's programme. On a heavily stocked shoot the area will be, at best alive with pheasant scent or, at worst, endless fresh birds. A good dog will, if put on the line, follow it patiently, hardly lifting his head as birds rise about him and duly deliver the wounded one. It is a very rewarding sight to see him come popping over the fence with the runner, a lingering death averted and another to add to the bag.

FIELD TRIALS AND WORKING TESTS

Apart from a very few trials held on grouse moors in Scotland and the north of England, the season starts in earnest in October when almost every week-day sees a retriever trial. The qualifying open stakes are spread over the weeks until early in December when the Championship is held. Goldens have, twice since the war, won this Blue Riband of the Retriever world. Miss Jean Train, now Mrs Lumsden steered F.T. Ch. Treunair Cala to victory and Mrs Atkinson's F.T. Ch. Mazurka of Wynford won it and also ran second the following year. For two rather tense days the best retrievers of the season battle out the issue, walking up and driving, and the standard of work has risen over the years. It is not uncommon to find in the final run-off that the winner and placed dogs are all handled by professional handlers.

Field trials are indeed very professional affairs nowadays and it is an education to watch these dogs responding eagerly to the human help in which they have complete faith. The skill in making maximum use of the dog's natural abilities with the unobtrusive 'hand on the tiller' is partly a gift, not given to many, but mostly acquired through a thorough study of the job. Hence the domination of the professional.

It is, I think, a reflection on field trials, that a spectating shooting man is a *rara avis* on any field trial day. It is seldom now that you will find him in the 'spare dogs/spectators' line, looking for the sort of dog he hopes to use next shooting season To get to the top you must have a field trial dog. The shooting dog is only a second class citizen in the kennel. The situation is analagous to the pre-war point-to-point when you raced your hunter. Now you hunt your race horse sufficiently to qualify him. The average shooting man is not a dog handler by nature or experience, but would not dream of shooting without this item of equipment. Should he buy a field trial winner the chances are that in a very short time the dog will be out of hand and unsteady; the model dog you watched apparently unmoved as birds dropped about him or a hare jumped up as he hunted for a pheasant, will, most probably have achieved the final indignity of a tether before the end of the season.

Inevitably in the public mind only those kennels which consistently run dogs in trials, are accorded the acclaim of 'workers'. There are, however pockets of unpublicised Goldens all over the country which fulfil shooting requirements generation after generation. Those breeders eschew the performance in public which sets the seal of worth for so many, and some are very critical of the value of field trial achievement.

To be a member of the field trial fraternity is one of the most enjoyable sides of gundog training and owning. Old friends and new gather at the headquarters and the pre-trial evenings are always very cheerful and convivial affairs. No one is kinder or more helpful to the novice than the 'pros'. Almost without exception they will give friendly advice and encouragement to newcomers.

One of the charms (and limitations) of field trials is the luck element. A draw and a ballot for places made some time before the day decides who gets a run and all qualifying stakes are at present over-subscribed. Should you be lucky and draw a number that number decides where you and your dog are during the first stages of the trial. According to whether the trial is judged by three judges or two pairs of judges there are 6 or 4 dogs in the line at one time and as each is tried and comes out, the steward feeds in the rest in strictly numerical order. You can run in trial after trial and only get comparatively simple retrieves, conversely you may get a near impossible job early in the proceedings or more sadly just when your hopes run high. You may not be in the line when a difficult runner is shot or another dog fails to find a bird and without demonstration of the ability to find these birds few open trials are won. Luck may and does decide your lot but without the ability to seize any opportunity offered a dog will not be in the money. Top class dogs are campaigned throughout the season, luck levels out and if he is good enough the dog will be there, poised for the highest honours of the Championship.

There are three Golden Retriever Clubs which run trials for Goldens ranking for qualification, as well as novice, puppy and non-winner stakes which are also confined, for the most part, to the breed. These minor stakes are a good starter, both for novice dogs and first-time handlers. Experienced judges take into account the age of the dog and status of the trial, and while no quarter is given or expected in open stakes, the standard is, of course, rather more relaxed in the minor stakes. A well trained shooting dog should have been schooled sufficiently to hold his own in one of these, when he has had some work and experience.

Of recent years gundog training clubs have sprung up all over the country, some independent of, and some attached to, F.T. Societies. Week-ends and summer evenings are the popular times for meeting and most people can now find a club in their neighbourhood. Young dogs can benefit from these classes; they see others at work and get used to gunfire and working on dummies while walking in line in the company of other dogs and handlers. Successful F.T. handlers are often available to help with problems.

These clubs often hold working tests. Dummies or dead pigeons are used and blanks are fired. It is doubtful if these have, in any way, contributed to progress in field trials, but it is a form of competition in its own right to anyone wishing to train a dog to the present high standard.

Tests at the Country Landowners' Association Game Fairs and country fairs always attract a large crowd, and very often provide an opportunity for the public to see top class field trial dogs in action, as the committee usually invite a dozen or so dogs to compete which have won trials the previous season.

The tests are laid out by field trial handlers, and judged by 'A' category judges, and designed to demonstrate the value of good marking, good handling and the finer points of training. The spontaneous applause which greets a smart retrieve completed or a temptation resisted, clearly shows how many folk appreciate a well trained dog. A running commentary by an expert is also very popular. Tests are a spectator sport while field trials are often the reverse as much of the work can be seen only from the line itself and is often out of sight of everyone else. Tests, on the other hand, can usually be watched throughout from one spot and often in good weather.

It is, I think, putting the cart before the horse to regard tests as a *preliminary* to trials. There have been examples of this in recent years when successful test dogs have been entered for trials with absolutely no experience of shooting. The result is often disastrous and must be a humiliating and bitter experience for the handlers. Perhaps in time these activities will go their separate ways; eventually test championships may be held as the ultimate for these dogs and handlers who specialize in this activity. So far, except that registration is a pre-requisite of entry with the clubs, the Kennel Club have not been involved in the test world. It is possible that some control will become necessary. I have seen tests laid out by enthusiasts which more closely resembled circus acts than any probable situation in the field. Maybe 'course builders' with experience of the real thing will eventually be considered necessary. Although still unpopular and greatly criticized by quite a number of influential members of the trialling world, I feel that training clubs and working tests have come to stay. For a variety of reasons shooting and field trials make no appeal to some and I know at least two owners who are definitely opposed to field sports and yet who, like their dogs, are enthusiastic supporters of a training club.

Goldens, perhaps because of their extrovert natures, have done extremely well in the Game Fair tests, winning them many times, and their obvious enjoyment and enthusiasm is very attractive. For speed and style the Golden Retriever bred for work, is second to no other variety.

GOLDEN RETRIEVERS IN GENERAL

The show Golden of today is a winner! The pups are amongst the most appealing of all breeds and sell readily. The breed has become very standardized. The leading kennels consistently produce very lovely animals conforming to the blue-print laid down for them. Show judges are appointed by the specialist clubs ensuring continued progress.

Very few people have the interest or knowledge to breed for work. Still fewer the inclination or opportunity to use their animals for the purpose for which, years ago, they were exclusively evolved. It serves no purpose to recriminate from either side or to fail to recognise that this situation has arrived. There are now virtually two varieties of Golden and the vast majority are show-bred.

It is quite impossible to know what has happened to the qualities required for a worker in these dogs. Good mouths, courage in cover and water, good noses, temperament and trainability are factors inseparable from the worker—none of these is known or required in the show ring. How much has been sacrificed on the altar of appearance no one can say, and for anyone with an established strain of workers it could be very costly indeed to find out.

I have been reproached many times for scepticism about breeding 'dual-purpose' dogs. The passing of the years has strengthened my conviction that it is not a practical proposition. Talking many years ago to the late H. S. Lloyd, whose achievements with cockers is legendary, I have never forgotten his answer to this question. 'No man can rise to the top of two professions.' The message is clear to us lesser mortals.

Watching the huge classes of Goldens at Crufts, one is very impressed with the level type of these leading dogs. A contrast to the early days when, pressed to show your workers, there were all sorts and sizes in the ring. As you admire the class winner you wonder what response you would get if you asked him to plunge into a weedy pond repeatedly in the dark as you finish a long day's shooting with an evening flight. Palpably unfit and over-weight, he is pretty obviously newly washed—a heinous offence to a gundog, destroying as it does the natural oily protection of the coat. He would probably get pneumonia if he got as far as the pond! A Golden can return from a long session of swimming in icy water and, if you turn back his upper coat, you will find his skin is perfectly dry. I often see Goldens sitting wet and alert at a drive when other breeds are shivering—maybe a legacy from the far-off Tweed Water Spaniel. Kept lean and fit Goldens last for many years and are going well at 10, 11, and 12 years. Their mechanics are right for the job and they live out a useful life doing what they most enjoy.

When in line with my husband on a covert day I am quite often told 'Aren't you lucky your dogs are so calm, our's has to be tied up he's so keen' or 'aren't you lucky your dogs don't whine with excitement'. Once a fellow wife said 'I'd love to have a dog sit quietly through a drive'. Then she bought a pup from a good kennel and trained it, since when she has won a few trials and tells me she is now often told 'Aren't you lucky'.

An eminent politician had a dog he adored but which whined and yelped through a drive. He had to blind-fold it in a grouse butt on pain of being forbidden to bring it, such was the noise. In despair he approached an old Scots keeper and asked him what had gone wrong. The sage pondered and then remarked 'An excitable mon makes an excitable dog'. Suddenly everyone in earshot was busily occupied with laces and equipment!

I feel I have indeed been lucky, lucky to have almost completed half a century of unbroken ownership of Goldens. Lucky to be still learning with every young one more about communication which is the basis of training, and with every day's shooting something more about their skills and instincts.

Although I do not subscribe to the anthropomorphic approach, I am deeply grateful for all the pleasure and happiness Goldens have given me and my family.

3 Flatcoat Retriever

Prior to the publication of my book, *A Review of the Flatcoated Retriever* in 1968 there was very little concerted writing about the Flatcoat and this was scattered in various books and periodicals published before World War Two. After this the absence of authoritative information led non-informed writers to surmise for themselves and distort the facts. Unfortunately, the one man who did write with real authority on the breed, the late Stanley O'Neill, wrote mainly in many personal letters and as an able columnist in *Dog World*. He had an intimate knowledge of the Flatcoat and was a close associate of many patrons and breeders for over 50 years and it has not been easy to amass and pass on his knowledge.

The complex origin of the Flatcoat, like that of the other modern British retrievers, is buried in obscurity and cannot be satisfactorily unravelled. Their beginning was the mixing of 'blood' of the various dogs used for sport long before the concept of the pedigree or show dog. In fact, until the middle of the 19th century, before which other dogs were recognised as pure breeds, the wavy and flatcoated retrievers were classed as mongrels.

The need for specialist retrieving dogs had arisen as a result of the changed character of the sport of fowling. Hawking and netting had given way to the increasing use of the shotgun and the development of shooting flying. This resulted in wounded birds being difficult to locate. The dogs available were spaniels, setters, water dogs, sheepdogs and hounds. The first main crossing seems to have been between setters and sheepdogs, the former for its exquisite scenting ability and the latter for its intelligence, biddability and retrieving instincts. A small infusion of hound 'blood' was also probably used to improve scenting power, the obvious choice being the bloodhound.

Setting dogs, mainly of spaniel origin, had been present in Britain for many centuries and had been much improved over the years. Many families and different strains developed in various parts of Britain with distinctive characters and colours. The latter ranged from black to red and parti-colours with various combinations of black, white, tan, liver, etc. We can only surmise as to which strains went into the making of the early wavy or flatcoated retrievers, but Edward Laverick (1800–1877), an ardent sportsman, had the leading strains for over half a century. These were very popular and would be the obvious choice before the Newfoundland cross was made about 1840. Sheepdogs available for the crossing were rough and smooth-haired collies which had been established for a very long time in Britain.

In a personal letter O'Neill gives his opinion that just as the prototype of the Curly-coated retriever appeared about 1800 as a result of the crossing of the new retrieving dogs with a poodle-like waterdog from our fishing boats, also long in use in Britain, so were British retrieving dogs later crossed with a so-called Newfoundland dog and the wavy or flatcoated retriever was the result. They were first called Labrador retrievers. They were mostly black with white and tan and some brindle.

It is a fact that there is endless confusion in the literature of the 19th century concerning the use of the description Labrador applied to dogs used in the making of modern retriever breeds. Although a direct descent from an original Labrador dog is postulated for the modern Labrador retriever, first registered at the Kennel Club in 1903, it seems much more likely that he originated subsequent to the Newfoundland cross mainly by selection of the smooth-coated puppies that were common in the litters of wavy-coats, with possibly minor infusions of other 'blood', such as that of the black pointers which were kept in distinguished retriever kennels. In the Victorian era foreign things were very much in vogue and there seems little doubt that some of the so-called imported Labradors were really bred in Britain and were taken to the ports of Bristol and Poole to be sold for high prices as imported Labradors.

The first Kennel Club register of dogs covering the years 1859–74 listed 210 retrievers. Of these 47 were curly-coats and the rest wavy-coats with one being described as flat-coated. Where the colour was given it was black or liver, but in many cases it was unspecified, possibly because the sought-after colour was black. By 1876 Wavy and Curly-coats were separately classified and by 1880 the colour of every dog was given. The great majority of Wavy-coats were black, but there were a few livers, one black and tan and one brindle in the volumes up to 1893. Breeders bred selectively to get rid of white and brown, and yellow-coloured puppies which appeared as recessives were very quickly disposed of; however, some people undoubtedly took a fancy to these and bred for this colour thus contributing towards the foundation of the Golden retriever. Until 1913 the latter was classified in the stud book with the liver-coloured Flatcoats whose hue had a wide range. Very occasionally a yellow puppy still appears in a litter of Flatcoats. The author bred one in 1962 from black parentage although a grandsire was a liver-coloured dog.

The great name in the history of the Flatcoat is that of Mr S. E. Shirley, 1844–1904. He was a most important patron of all kinds of dogs and a keen sportsman. He founded the Kennel Club in 1873 and became its first president and chairman. He lived at Ettington Park near Stratford-upon-Avon and was a member of Parliament. He was a show and field trial judge.

Shirley's father kept a large kennel of retrievers long before the days of the show bench, and, as a youth he became very interested in them and determined to improve the breed. The neighbouring gentry also had strains of their own which he used.

What he did was to breed together 'Labradors' that had the Newfoundland cross and were mainly solid black and called them Wavy-coated retrievers. Croxton-Smith, a reliable authority, wrote in 1950 that Shirley began by trying to get rid of the conspicuous faults attributable to the Newfoundland and to get a flat coat instead of a wavy one. In order to do this it was supposed that he used Collie 'blood'. The conspicuous faults of the Newfoundland were coarseness and colours other than black.

Rawden Lee (writing in 1893) denies that Shirley used any further setter cross, but states that he bred by careful selection from old local strains to a stable type. In an interesting letter that appeared in the *Kennel Gazette* of April 1880 Shirley described the characters and wins of a number of individual dogs belonging to various owners and placed them in three categories, the Labrador, the Setter and Collie types. So it appears that what Shirley accomplished was a skilful integration of these into one standard type. It was a dog that took sportsmen by storm and became the most popular and fashionable sporting dog around the turn of the last century.

Many of the early show-winning Flatcoats which are listed in the first Kennel Club stud book are featured in the pedigree of Zelstone, a considered pillar of the breed. He was the sire of Ch. Moonstone, a dog said to have been unbeaten on the bench and whose daughter Donna was the dam of Ch. Darenth from whom all present day Flatcoats stem. Darenth's sire, Ch. Hopeful, also brought in many of the early celebrated dogs. Darenth belonged to the noted Kent cricketer Mr Alan Shuter, for whom he earned a small fortune at stud. Another very celebrated dog, a great grandson of Ch. Darenth was Col. Legh's Ch. High Legh Blarney for whom Mr H. R. Cooke paid 200 guineas in 1905. He also was a valuable stud and the sire of Ch. Jimmy of Riverside and Ch. Longshaw Bruce representing two of the three male lines of descent which continue today. The third line comes through Ch. Wrangler and direct descendents of dogs bearing the 'Black' prefix of Major Harding Cox.

Cox and Cooke were two other notable sportsmen who were contemporary with Shirley and helped to perfect the Flatcoat. The former bred Black Drake whose success at stud was phenomenal. He also wrote a number of sporting books in a loose and breezy style in which he told of the exploits of his Flatcoats and his great love for the breed as a companion and excellent gundog, but the substance of his article in Hutchinson's Encyclopaedia in about 1934 is vague and sketchy and gives a mis-leading description of the Flatcoat, giving him a black setter image. On this subject O'Neill wrote: 'It has been written that a Flatcoat resembles a black Irish setter. Except that both have featherings on the legs it is difficult to find any point of resemblance and it cannot be too strongly emphasised that this statement was made in a work designed to appeal to the general uniformed public, and its purpose was to give those who had no knowledge, the most striking and easy form of recognition.'

Mr H. R. Cooke, 1860–1951, was a most outstanding patron of the Flatcoat, being closely associated with the breed for 75 years. He saw it through its hey-day, followed its fluctuating fortunes through wars and peace and witnessed its failure to rise significantly in numbers up to the time of his death. He has left accurate records of all his 'Riverside' dogs. These include 233 pedigrees and many columns of press-cuttings relating to field trial reports and show critiques. Eighty-seven Flatcoats were home-bred, but 146, many of which were sired by his own dogs, came from outside kennels. This remarkable interest led to a fabulous kennel, one that is possibly unique for any breed of dogs. In 60 years of showing he won 3,600 high awards and special prizes in breed and any variety competition and made up many champions. He used his dogs in the field throughout his life. None were kept specifically for trials, but those he ran were worked in the normal way. He won about 60 awards in field trials and made up two field trial champions, Grouse and Toby of Riverside.

From time to time he deplored in the press that not enough owners worked their dogs and he expressed the fear that the purely show interest might overtake the breed. He was an advocate of a medium-sized dog as being best for work, by which he meant, I believe, a dog of about 25 in. at the withers.

After the First World War, following an initial rise of Kennel Club annual registrations of Flatcoats up to 438 in 1924, there began a downward trend that continued until the onset of and during the Second World War. The number in 1938 was only 110 and this figure was not reached again until ten years later.

Although the Flatcoat was originally the dog of wealthy sporting patrons, a number of gamekeepers became strongly attached to it during the inter-war period. They recognised quality and bred for this as well as for working ability. They included W. Skerry (Towerwood); J. Inman (Danehurst); Ellis Ashton (Leecroft); Adam Gordon (Blackdale); W. J. Simms, A. Southam, Colin Wells, E. W. Bryant (Ponsbourne), J. Boyd in Scotland and others.

Cooke's Riverside kennel drew heavily upon the keepers and other breeders and its pre-eminence discouraged many breeders who gave up, so by the middle of the thirties there were only two main breeding kennels. These were Mr Southam's and Mr W. J. Phizacklea's (Atherbram). The former kept good quality bitches and used the best sires available. His Ch. Specialist ran well in trials and was the sire of F.T. Ch. Windle

9. The Flatcoat is traditionally the gamekeeper's dog. Mr Colin Wells with his Ch. Wave.

Popular. He gave up breeding with the onset of the war, but Mrs Hemm acquired some of his 'blood' and carried it through the war, but, unfortunately, the male line of descent through Ch. Specialist did not continue, though the distaff side continues today via the (Pewcroft) breeding of the late Stanley O'Neill.

The lines of the other main kennel, Atherbram, rested on a broader basis. An early notable bitch, A. Biddy, was the dam of F.T. Ch. Windle Popular and of the outstanding bitch Ch. Atherbram Jet who produced several brilliant progeny in three litters by Ch. Dancer of Riverside.

An outcross bitch, Rettendon Dido, brought in lines from Southam's dogs and Dual Ch. Toby-of-Riverside. Phizacklea also bought Toby-of-the-Hill, a grandson of Dual Ch. Toby of Riverside, who became the grandsire of Mr Colin Wells's Waterman, a noted champion and field trial winner.

During the Second World War Phizacklea acquired a bitch fortuitously. She was Cemlyn, the dam of the author's well-known Claverdon Jet, a champion and field trial winner. Cemlyn came from very good field trial stock. Her sire, Windle Don, lined back to F.T. Ch. Elwy Mary, a daughter of Ch. Dandie of Shipton and Bibby, both field trial winners, and on his dam's side to Maes Judy, a daughter of the Dual Ch. Toby-Bibby mating. Cemlyn's dam, Windle Peggy was also a F.T.W. and was the daughter of F.T. Ch. Windle Popular by the consistent trial-winning dog, Quick of Riverside. It was unfortunate that, although Phizacklea was a sporting farmer and used his dogs in the shooting field, he had little interest in field trials. A black visiting bitch, Roland Peggy, bred to one of Phizacklea's stud dogs, Atherbram Nimrod, produced two liver puppies, a dog, A. Rufus and a bitch, A. Bess. He became interested in this colour, and, by breeding fairly closely, developed the makings of a pure liver line that his wife continued after his death. Mrs Barwise, (Forestholm), also used A. Rufus and subsequently bred some liver Flatcoats. She also bought a recessive liver bitch of black parentage from Mr Sydney Guy (of Sauch), and these various liver Flatcoats were used by Mrs Izzard, (Ryshot), in the foundation of her present pure strain of livers.

So it was that the majority of Flatcoats available in the late 'forties and 'fifties were from the Atherbram kennel. Other patrons who helped to keep lines during the war were Mrs Barwise, Mrs Hemm, (Adlington), Miss Meeson, (Rettendon), Mr F. Allen, (Peddars), Mr J. Spencer, (Overash), Mr R. E. Birch, (Brynasaph) and Mr Cooke, although the latter let a number of his Riverside lines lapse. Towards the end they were joined by the author, (Claverdon) whose foundation stock was Ch. Claverdon Jet and Revival of Ettington who lined back to Ch. Specialist and Leecroft Buxton, and, after return from service duty, Mr O'Neill, Major H. A. Wilson, (Nesfield), Capt. Wilson Stephens, (Hartshorn), Mr Colin Wells and other keepers including E. W. Bryant, (Ponsbourne), J. W. Moore, (Daleland), K. Barber, (Denmere), W. Allen, W. Worby, F. Craner, E. Kennett, A. V. Lee, P. R. Standley and J. Boyd in Scotland. All these people provided contributions to Flatcoat recovery.

Stanley O'Neill must be credited with the most valuable service in picking up the threads of disappearing lines. Inconvenience and cost to himself and his wife were put aside unselfishly. There was no question of breeding to type. Every available dog had to be sought and he gave constructive advice to many breeders. He was completely devoted to his Flatcoats and used them all regularly for rough shooting. He considered that there was no better gundog as a maid-of-all-work, but he had little liking for field trials.

He had kept a very good bitch, Pewcroft Peg, stemming back to Ch. Specialist on her dam's side and by P. Puzzler, one of the memorable 'nick' between Ch. Dancer of Riverside and Ch. Atherbram Jet. He bred her twice, firstly to Bryn of Adlington, bringing Specialist in again. In this litter were the brother and sister champions Plug and Pitch. Secondly Peg was bred to Sweep of Riverside, a grandson of both Ch. Specialist and Ch. Dancer of Riverside. Sweep's dam, Bryn Asaph Fair Trade, had excellent field trial potential behind her mainly from the Bibby matings to Ch. Dandie of Shipton and Dual Ch. Toby of Riverside.

This second litter produced some very useful stock including Major Wilson's Pewcroft Page who was an excellent trialer and the sire of the author's well-known Ch. Claverdon Jorrocks of Lilling, also a field trial winner and winner of the Country Landowners' Game Fair Retriever test in 1961. In the same litter were Claverdon Pewcroft Pieman, the sire of Claverdon Tawny Pipit and Claverdon Turtledove, winner of two trials, and the sisters P. Picture and P. Peep. The latter has played a significant part in shaping Flatcoat history through her mating to Mr Brian Farr's Gaff of Riverside by Nobbie of Riverside, who, with his had sister, Joy of Riverside, had several successes in field trials towards the end of Mr Cooke's life.

Some of Mr Birch's excellent working and field 'trial blood' has come down to us today through the Peg-Sweep mating and also through Sweep's two sisters, Waff and Wren of Riverside. The former was bred to Nobby of Riverside and the latter to his sire, Smoke of Riverside. However, other Bryn Asaph and some excellent Welsh working lines died out, but a useful outcross line came from Mr S. Guy's Patch and Dot of Sauch, both of whom were line bred to F.T. Ch. Elwy Mary through Brek, their dam. This line comes to us today through Mr Barber's Denmere Prince and his sister Judy of Sauch. O'Neill used Prince and his son Denmere Bruce and provided some very useful foundation bitches for today's Collyers, Fenrivers, Blakeholme and Yarlaw kennels and a dog, P. Prefect went to Mr Homer Downing in America and began a revival of interest in the Flatcoat in that country, a revival that has since gone from strength to strength.

Soon afterwards Downing imported a liver bitch, Atherbram Stella. This kennel, Bramcroft, has specialised in obedience training, as have other American patrons and Flatcoats have done particularly well in this sphere. Downing wrote in 1972; 'It has been about 20 years since Flatcoats made their debut in Obedience in the United States. They really were a smashing success individually and collectively. The eight in the Ohio area had but one non-qualifying score among them in gaining their titles—unbelievable!'

10. Mr P. Johnson's Flatcoat dog Ch. Woodway, a powerful swimmer.
Lars Söderbom

After her death in 1956 Mrs Hemm's line carrying Southam's 'blood' with Revival of Ettington as an outcross, was continued by the two keepers Moore and Boyd. The latter bought a son of Peddars Flip by a brother of Nobby of Riverside called Claxton Cadet. This dog lived up to his name as he was of a small light build. Boyd bred him to Zoe of Adlington and to another bitch of Sauch-Riverside origin and subsequently brought the two lines together. The resulting progeny were used by Boyd, T. R. Naisby and other sportsmen in the North who found them to be excellent tireless workers on the grouse moors. Naisby told the author that no amount of work caused them to suffer from heather burns as their legs were protected by their feathering.

E. W. Bryant obtained his foundation bitch from Skerry whose dog, Ch. Towerwood Vigour was a direct male descendent of Ellis Ashton's Ch. Longshaw Bruce. Vigour was the force behind two lines of field trial winning Bryn Asaph and Riverside dogs. Bryant also used some Atherbram and Bryn Asaph 'blood', the latter through Ponsbourne Peter whose line is continued today.

Writing on field trials in the Labrador Retriever Club Stud Book and Record of Field Trials published in 1949, Mr Mackay Sanderson gives a detailed account of Flatcoat field trial history tracing it back from the first International Gundog League retriever trial in 1899 right up to 1948 when Flatcoat field trial activity was very small. He mentions outstanding dogs either as competitors or stud forces in relation to working ability. These included Ch. Darenth, his son Horton Rector and grandsons, Sweep of Glendaruel and Ch. High Legh Blarney, Bergholt Dale, Ch. Wimpole Peter, Pitchford Marshall, Ch. Royal River and Dual Ch. Grouse of Riverside. He made a special reference to the bitch, Meeru who ran extremely well in trials against heavy Labrador odds. She won the Retriever Championship stake in 1912 and was second in 1913.

Sanderson states that during the interwar period 1920–38 only about 70 Flatcoats competed in trials. None were placed in the championship stake, but diplomas were won by Ch. Dandie of Shipton, Bryn Asaph Bob, Elwy Rita, Elwy Doris, Windle Peggy and F.T. Ch. Windle Popular.

The patrons who sponsored the Flatcoat in trials during this period and to whom the breed owes permanent indebtedness include Lady Wigan, Mrs E. G. Oliver, Mrs W. Fletcher, Mr R. E. Birch, Col. Winch, Col. Elwes, Mr Muir Wilson, Mr J. H. Hulme, Mr P. Barret and Mr R. Sharpe.

Sanderson gave Mr Birch credit for the greatest of field trial wins with the dogs F.T. Ch. Elwy Mary, Bryn Asaph Wyn, Jean, Soloman, Adam, Beck and Fair Trade. He cited the bitch Bibby, bred by Mr F. Perry in 1921 and owned by Mr Barret as the greatest matron where work was concerned.

CHARACTER AND WORKING ABILITY

The Flatcoat is a dog of distinct character that many people who have lived with him agree is inimitable among dogs of any breed, for he has a subtle, indefinable charm. He is a confirmed tail-wagger who never sulks or resents correction, but has a way of getting back into favour very quickly. His temperament is gay and he is always ready for fun or sport. In the matter of growing up he resembles a canine Peter Pan, something which can be of a disadvantage in his training programme.

He has a high intelligence and sagacity and is a very talented worker; moreover he is a handsome fellow. A writer in *The Times* (of London) in 1958 described him very aptly as an Edwardian whose grace and distinction matched that of the epoch in which he had his hey-day. This writer was of the opinion that his decline in favour was partly due to a gayness in youth that is combined with a slowness to mature and to take his responsibilities seriously, and partly to a monopolisation of the best dogs at one period, by a few wealthy patrons, who inflated prices, making the market prohibitive for others.

The Flatcoat has an inherent natural ability to hunt and find game and to retrieve it to hand. The retrieving instinct is so marked that the only guidance needed in this respect is the perfection of the style of delivery. Neither is there any better gamefinder whether the scent is hot or cold. He hunts with eager drive and the greatest of perseverance. He is not put off by nettles, thistles, the thickest cover nor the coldest water. His intelligence in respect of marking develops quickly in the field and the recognition of wounded birds in flight which fall far behind the line becomes second nature and these are invariably brought to hand in due course. Hard mouth is uncommon; in fact, it is exceptional.

It is not surprising that once an owner has experienced the pleasure given by this delightful dog he invariably retains his allegiance to the breed. When asked about this one sportsman wrote to the author;

'I prefer the Flatcoated retriever to any other variety because, at a formal shoot he will do all that the other varieties will do, and is a quicker worker, better in very cold water and bolder when looking for a runner in very thick covert. For rough shooting he is very easily trained to point his game, in fact, he does so naturally, and, if properly handled, soon becomes as staunch as a good setter. I find this very valuable when walking round alone with the gun. I have also found that he takes a little more readily to guarding anything I leave him to mind. Some people may not agree with me entirely, but this has been my experience for over 50 years.'

Another sportsman who has used Flatcoats regularly for many years in the shooting field and whose criteria of a good retriever are obedience, courage and determination in covert and water, good stamina, speed in working, good nose, soft mouth, complete steadiness and a good conformation writes: 'I have plenty of scope for testing their characters as there are a lot of brambles, bracken, sugar-beet and water here and at most places where I shoot. Also, during the last month of the shooting season, I use them on the bracken banks for pointing cocks, which they do most effectively and steadily.' Of the characteristics of the breed he says; 'Having had Flatcoats for some twenty odd years now I have come to the conclusion that there can be very few trainers left of the 'Bill Tansey' stamp, who understood the breed so well.' (Bill Tansey was the last of Mr Cooke' gamekeepers and he just loved his Flatcoats.)

Flatcoats are now well entrenched in 'hunting' circles in the United States of America and, to a lesser extent, in Canada. The North-Americans favour large dogs because their cover is commonly heavy.

The Flatcoated Retriever Society of America has recently published a number of reports from sportsmen commenting on the Flatcoat as a 'hunting dog'. The following from John S. Jacob expresses the thoughts and feelings common to others: 'I prefer the Flatcoat to any other field dog basically because of their size and because they are not afraid to go in any brush or sloppy water. They are an extremely strong dog with good spirit and excellent stamina. I wouldn't use another dog in the field. Other people to whom I have spoken feel basically the same way. I have never heard an owner say that he was dissatisfied with a Flatcoat's performance and ability. Trainers I have talked to have commented on the ease of their training and the willingness of the dogs to learn and respond to training.'

I was amused at the story of the Flatcoat that earned quite a lot of money for his owner by swimming up and down the banks of the Missisippi spotting mink dens.

In this country, compared with Labradors, Flatcoats are favoured in the shooting field by a relatively small number of sporting patrons. These include Brigadier Clowes, Mr B. Farr, Lord Roborough, the Earl of Devon, The Rev. J. W. Steele, Lord Brook, the Lord Jeffreys, Mr D. Jessel, Mr Vesy-Holt and others. They are also used by the dual-purpose breeders whose interest in fieldwork is in 'picking-up' and trials. These include Air Commodore W. Hutton (Yarlaw), the Hon. Mrs D. Jessel (Collyers), Mr Read Flowers (Fenrivers), Mr and Mrs D. Izzard (Ryshot), Capt. Wilson Stephens (Hartshorn), Miss C. B. Hall, Mrs J. Mason

Heronsflight), Mr and Mrs P. Forster, (Wizardwood), Mrs J. Shore and the author.

Many gamekeepers such as Colin Wells, Warren Wells, Peter Johnson, (Downstream), J. Boyd have held an unswerving devotion to the Flatcoat and many others are now realising his worth in their everyday multifarious needs.

TRAINING

A Flatcoat should preferably be kennelled from an early age. If he is given a comfortable bed in dry, draught-proof conditions he can withstand a deal of cold. Too much licence leads to boredom and reduces pleasure and concentration during training sessions, and, most important, it allows the young dog to indulge in misdemeanours. If kept in and around the house the owner cannot concentrate on the dog and so he pleases himself and not the owner.

However, it is a good plan to have the dog in the house or part of the house sometimes when the owner can afford the time to concentrate on it, and, during that time to insist on implicit obedience. The puppy should also be taken about in the car, meet the world in general, be familarised with cats, dogs, hens and farm animals and here learn the meaning of the word 'No!'

A daily grooming removes dead hair and dirt from the coat and is conducive to a good coat condition. It affords an opportunity to appraise the dog's general condition and to look for injuries or other skin lesions. A working dog should be lean and in good muscular tone. The coat is water-resistant and dries quickly; in this matter its length is a decided factor in dispelling water on shaking, an advantage not present in short-haired dogs. Eyes should also be inspected for the presence of foreign bodies such as grass seeds collected during work, as they act as irritants.

Like children, young puppies can be trained in a small way from the age of 8 weeks onwards, but the beginning of field training is best left until the age of 7–9 months and then it should be a gradual process. Short lessons, constantly repeated, are the keynote of success, but the dog must never be allowed to become bored because concentration on the part of the animal is vital; of course, it is equally important that the handler is in thc right mood to concentrate on the dog.

Punishment for disobedience must vary from a scolding voice to a thorough shaking by the neck according to individual sensitivity, but it is imperative to administer this during the act so that there is no possibility of a misunderstanding in the dog's mind.

Whether a dog is to be worked only in the shooting field or also trialled, the basic need in training is for strict discipline and obedience. This, of course, is fundamental to all gundogs, but is particularly necessary for the Flatcoat because he is self-reliant and likes to work independently. This can be an advantage sometimes and is linked to perseverance, but, at times, in field trials and in the ordinary shooting field, the dog must work under guidance from his handler and give instant obedience to whistle and hand signals, although it should be borne in mind that there is a danger in overdoing these, and so interfering with the dog's natural ability to find game.

The training ground should be varied as much as possible as strange ground with its new scents is always an exciting factor. This applies to water as well as land. Puppies should be introduced to the former in the summer months when swimming is a greater pleasure. The dog should not be taken into the shooting field until he can be relied upon to behave himself and then only for short periods on small days. In the beginning it is very unwise for the handler to carry a gun himself because he must be free in his mind to concentrate on the dog in this exciting atmosphere in which steadiness can easily lapse. Any tendency to squeak or whine from excitement must be checked immediately otherwise whining may become a habit.

Flatcoats have plenty of drive and some can be headstrong. Their well-developed natural working instincts which cause them to want to work independently make them concentrate so much on game finding that, although commands and whistles impinge upon their ears, they are not always transmitted to their brains and they go unheeded. Hence the necessity of implicit obedience in their preliminary training. A would-be successful trainer must have a very firm attitude of mind towards a Flatcoat. Ladies are nearly always too soft but he can also tax the ingenuity of a male trainer. He is certainly not the dog for a tyro, but a challenge to the best of trainers.

PRESENT DAY FLATCOATS

From what I have written it will be realised that the Flatcoat was a veritable dual-purpose dog as soon as shows were initiated because they began as competitions among working dogs. The breed has kept its dual-purpose role much longer than other gundog breeds. In this respect its fall from popularity has been a godsend to the preservation of its working ability because the dog remained mainly in the hands of sportsmen and gamekeepers with whom it has maintained its association. Such owners did not hesitate to cull inferior or unsound animals and they bred selectively with a bias for characters associated with work and stamina.

Popularity of a gundog breed invariably brings with it ill-planned and unnecessary breeding for profit alone because it creates a large pet market out of which unscrupulous breeders make capital. At the same time many have a complete disregard for the future well-being of the puppies many of which may find their way into unsuitable homes.

This state of affairs also pre-disposes to an increased incidence of hereditary disease as only responsible breeders will take the necessary steps to eliminate these faults from their stock. On this subject L. C. James

11. Mrs S. Hogstedt's Int. Ch. Claverdon Fantasia, a winner of two Field Trials.

12. A Flatcoat enjoying a field day.
John Marchington

writing in 'Our Dogs' (May 1973), points out that up to 30 years ago there was no scientific basis for dog breeding. It was carried out on pedigrees and the personal knowledge of the conformation, soundness, character and performance of the dogs themselves.

Sound working Flatcoats have been produced for over 100 years without radiography. The criteria of soundness have been trueness of movement and the possession of sufficient stamina to stand up to hard work. Dogs that did not come up to standard on these scores were discarded as breeders.

It could be that because of this selective breeding, although Flatcoat breeding has been based narrowly for very many years, with most families showing inter-relationship, the incidence of hereditary disease appears to be appreciably lower than in other gundog breeds as judged by American, European and British figures available at present for the incidence of hip-dysplasia.

In gundog breeding the important thing is selection for soundness and stamina; the latter cannot be expressed unless the dog is properly worked. In this respect the Flatcoat has been lucky for much of his existence, but this luck cannot hold because over the last decade there has been an upsurge in his popularity and he is getting into the hands of people whose only interest is in showing him and who therefore cannot selectively breed to maintain working ability. So, as a working dog he appears to be at the cross-roads and looks like going the same way as the other gundog breeds. Several sportsman breeders have affirmed that popularity and the show fancy have already wreaked a loss to the working potential of the breed and that those who wish to improve this to the standard of the best working strains of other retriever breeds in field trial competition will have

13. The author's Flatcoat bitch, Claverdon Turtle-dove with Field Trial trophies. Winner of two Field Trials.
S. Jay

to drop their long held dual-purpose conception and selectively breed towards this end.

The annual registrations of Flatcoats at the Kennel Club are now running at about 300 per annum. The present Breed club, Flatcoated Retriever Society, has more than 300 members under the presidency of Brigadier Clowes. It sponsors two annual field trials, a 12 dog Open Qualifying stake and a Non-Winner stake, both for Flatcoats only. It also holds an annual working test and issues a comprehensive Year Book containing records of field trial and show wins, news of Flatcoats in Europe and America and original articles. The Hon. Secretary is the Hon. Mrs D. Jessel.

The Society field trials have been well patronised by a nucleus of members since Mr Birch organised the first post-war trial in 1948, but they would benefit by additional support.

A number of patrons has been successful in Any Variety trials. These include the Hon. Mrs Jessel with Claverdon Skipper, Ch. Collyers Blakeholme Brewster and Collyers Skeets; Air Cdre. Hutton with Ch. Pewcroft Prop of Yarlaw, Mr P. Johnson with Keeper's Girl of Chadwell, Mr E. O. Baldwin with Claverdon Fabulous and Claverdon Rhapsody and the author with Ch. Claverdon Jet, Ch. Claverdon Jorrocks of Lilling and Claverdon Gaff.

Colin Wells has always kept the dual-purpose image in mind and has produced many bench champions, some of which have been successful

1. A Yellow Labrador picking-up on a grouse moor.
G. L. Carlisle

2. A Golden Retriever produces a cock pheasant from the brambles.

3. A consistent bag-filler in the field.
Dr. N. Laughton's Flatcoat dog Rungles Jerome.

4. Curly-coated Retriever.

in breed and Any Variety trials, but all have had to work hard during the shooting season under conditions not always conducive to putting on the polish and control which put a good dog to the top in trials.

Major H. A. Wilson is the most successful trainer-handler. His kennel was re-stocked after his war service, with Claverdon and Pewcroft 'blood' under the Nesfield prefix. His dog Pewcroft Page was a great trialler in the few stakes in which he ran. Subsequently Wilson won many awards in Any Variety stakes with a number of his Flatcoats, and he has made up two field trial champions, the home-bred dog, F.T. Ch. Nesfield Michael, by Blakeholme Joiner and International F.T. Ch. Hartshorn Sorrel who was bred by Capt. Wilson Stephens out of a Nesfield bitch and sired by Mr Farr's Teal of Hawk's Nest.

Sorrel won the Flatcoated Retriever Society's Open stake in 1965 and the Ulster Retriever Club 24–dog qualifying stake two years running, in 1966 and 1967, as well as the Irish Retriever Club's Ch. stake in 1966, in which year she won a Diploma of Merit in the Retriever Ch. stake. She continued her successes by winning three open stakes in Ireland in 1969.

It is highly likely that Major Wilson's outstanding successes with Flatcoats may be attributed to his thorough understanding of the breed and an appreciation of the necessity for the strictest discipline in training them.

4 Curly-Coated Retriever

The Curly-coated Retriever is probably the least known at the present time of the four retrieving breeds that have been used in this country for well over a hundred years. Their job is to retrieve dead and wounded game only, and they are not to be confused with the continental pointer-cum-retriever breeds introduced to this country in recent years. These native retrievers are four quite distinct types of dog, each differing from the other in appearance. The curly is instantly recognisable by his coat, which looks like astrakhan and is a mass of close tight curls from the back of the skull to the tip of the tail. The coat which is either black or liver in colour is dense and though it has no undercoat, it is oily and consequently water proof. This impermeable quality is important in helping him to withstand the rigours of a long day's work in wet conditions which would reduce many other breeds to misery. The head however, is free from curl and is covered with short smooth hair, revealing the eyes, which are either black or very dark in colour, and have an intelligent expression. The eye should never be light in colour; in my opinion a light eye is detrimental to the expression of a gundog. The curly is a strongly built dog, standing about 27 in. at the shoulder and weighing between 70 and 80 lb., with a deep chest and good rib cage providing plenty of room for the heart and lungs of an active dog. Altogether the curly is of distinguished and handsome appearance. Being now uncommon he generally provokes interested comment from those who see one for the first time and speculations as to what kind of dog he is. These range from 'curly labrador' to 'some kind of poodle' or even plain doubt as to his parentage. Occasionally one does meet some one who recognises him. I have noticed this particularly in Scotland, where to judge from the number of exclamations of recognition, one might be tempted to think that nearly everyone's grandfather or uncle once owned one. There are I suspect more curlies up and down the country in general than one might think, and it would be interesting to know how many and particularly how many of them are worked as retrievers.

Certainly in their heyday about 70 odd years ago, before the Labrador as we now know it became the really popular retriever, curlies were well thought of and every keeper had several. They were bred by them as gundogs and as guards. One can well imagine that a curly would make a formidable companion to any keeper out after poachers, for he is invisible in the dark, his matt coat reflecting no light so that there would be no glint from it and even on a moonlight night he would blend well with the shadows. A strong eighty pound dog would fell a poacher with one spring. The curly's prowess as a retriever is praised in books of the period and one enthusiastic author recommends him as the one to be preferred to other retrievers, because, although 'being harder to train, is quicker and more dashing, and also more clever. Furthermore he looks dignified, his actions showing the height of animal intelligence, which at the same time being affectionate and companionable, keeping his affection for the master who owns, houses and works him. He often has to act on his own, and if well trained his sagacity rarely fails him'.

Such a dog must have an interesting history, but unfortunately his ancestry is wrapped in mystery. Little has been recorded and it is subject to a great deal of speculation. Retrievers as such came into being as a specialist gundog as a result of the introduction of the percussion shotgun in 1830 which changed the style of game shooting and brought about the

14. The Curly-coated Retriever.

15. A Curly, undaunted by a thorn hedge, coming through with a bird.

necessity for a dog that would bring back dead and wounded game from a distance, often out of sight of the gun. Writers of the period often refer to retrievers without being very specific as to what kind they were, though both curly coated and smooth coated are mentioned. It would appear that any dogs with good retrieving talents were mated without much attention being paid to their looks. Indeed it seems that a good deal of crossing of retrievers with setters, pointers and spaniels went on in an attempt to find the ideal retriever. A cross between a small retriever and an Irish water spaniel is particularly recommended by one writer as producing a very good dog, with a curled coat. This may have been an ancestor of the curly we know today. The small retriever may have been the St. John's type of Newfoundland dog, prized for its stamina in retrieving from the sea in that part of the world. The curly's coat has led to suggestions of poodle in its origin, and undoubtedly the poodle was much used for retrieving and tracking on the continent of Europe. Others have suggested that it is descended from an indigenous British dog known as the Rough Water dog, described as being intelligent, strongly made, curly coated, of aquatic habits and much used by wildfowlers. In Scotland there was, at one time a dog known as the Tweed Spaniel, a large curly coated dog and much used in retrieving from water. This too may well have played a part in the curly's origin, especially in view of all those curlies north of the border in the past. Though it is unlikely that we shall know now exactly how it was evolved, the curly came into its own about 1860 by which time the breech loading shotgun had become the weapon in general use. At this time there was a tremendous increase in game rearing and conservation and the big driven shoots became established as the general method of shooting, supported by retrievers and pickers-up. Bags of game were enormous and consequently so was the demand for retrievers. It is indicative of how great this demand was when one reads that one of the best shots of the period, Earl de Gray, later Marquis of

16. The dog sets off at speed with a retrieve from water.

Ripon between 1867 and 1923 had a personal total of 556,00 head of game. This was the period of the curly's greatest use and he remained actively on the shooting scene until the advent and rapid rise to popularity as a retriever of the modern Labrador. This newcomer had many things in its favour which the curly lacked. It was comparatively easy to train, matured more quickly, was a handy size, quickly regained its smart appearance after a day's work and did not retain a coatful of burrs like the curly, which necessitated hours of careful work to remove.

The introduction of dog-shows is also blamed for the wane of the curly's popularity as a worker, breeders being accused of producing beautiful but brainless dogs, and by the early 1900s writers were commenting on their poor showing at field trials. Certainly any gundog, that wants to earn respect as a working breed, must be able to put up a creditable performance at working tests and field trials before anyone is going to take the breed seriously. Sad to say, there is no curly-coated retriever on the field trialling scene at present, though that is not to say that the breed is no longer capable of producing such dogs.

The curly's temperament is kind and great hearted, as is usual in big dogs and totally without malice. He bears no resentment for scoldings to which he seems to have a philosophical attitude. He likes to be praised, is devoted to his owner and, if a family dog, to the other members of it. He has also quite definitely a sense of humour and loves a bit of fun and to play the clown. A favourite game is to put his head down and push with his forehead against one's knees or against the floor, reminiscent of the Lost Boy's defence against the wolves in Peter Pan. He is good with children who seem to like his cuddly coat. My own dog though not brought up with young children as my family was already older when we got him, has none the less always been instinctively gentle and tolerant of visiting children and those he meets on the street, being very willing to be patted and prodded. One has heard tales of curlies being virtual nannies to their owners' children, accompanying them devotedly in all their play and excursions and generally keeping an eye on them. One would imagine children being well guarded by such a companion.

However, for all his love of fun and companionable qualities, the well trained curly is a different character on the shooting field. At the sight of a gun and gamc-bag he at once assumes the professional air of one who knows his business and how to go about it. He walks quietly to heel of the man with the gun and does not take his eye off him. As a worker he is responsive, handleable, intelligent and has initiative and courage. He will face the toughest cover, while his strong physique enables him to work long and tirelessly in cold and wet conditions. In fact, to be in and out of water is what he revels in, being a naturally strong swimmer. He is said to be able to swim in 7 knot tide and to have been used at one time on the Norfolk barges to swim ashore with the painter. He will hunt under water for wounded game, diving and swimming below the surface for it, very useful when after pricked duck. He is therefore particularly favoured by wildfowlers, who appreciate his prowess in water. Often they prefer a liver-coloured dog as that colour blends better then black with the natural colour of the marsh lands. There is one estate in Scotland where curlies have always been kept and still are, for their particular value as water dogs.

They can of course be used on any game, be it fur or feather, like any other of the retrieving breeds. In one case, many years ago a curly was used for tracking and pulling down wounded roe deer in the forests of an estate. This particular specialisation however, made the dog quite useless for retrieving other game, according to the keeper whose dog it was, because it became unsteady and hard-mouthed.

Curlies have been accused of being hard-mouthed; this is not so. Curlies are no more hard-mouthed than other gundogs, though, like any, can become so given the circumstances. Any dog that has been injured by wounded game—a rake on the nose or eye from the spur of an adult cock pheasant for instance—will often tend to nip subsequent runners. Normally they are soft-mouthed and will efficiently bundle and gently carry a retrieve.

Assuming one wished to own a curly as a working gundog, how would one go about it? At once there is a problem. There is as far as one can ascertain only one kennel in this country at present breeding and raising curlies primarily for work. There are indeed not many breeders of curlies at all and in most cases their dogs are mainly for show or companion purposes. Of these there are one or two breeders some of whose dogs are worked. Incidentally curly bitches which have been spayed make good guide dogs for the blind and are regularly trained as such.

One can either acquire a trained dog or buy a puppy to raise and possibly train oneself. Should one be a novice trainer, there are several very good books available on the training of retrievers, which one would be well advised to read and inwardly digest. Novice or otherwise, one can learn a great deal from the recorded experience of others and various angles on the same subject are always interesting, for no two dogs or trainers either for that matter are alike. However, there are one or two points to bear in mind with regard to curlies and training them.

If what one requires is a retriever puppy to start gun training at 9 months old and be all ready to take out for his first season at about 15 months, then one should get a Labrador; a breed for which I have great affection and respect; but do not get a curly. For a curly is a slow maturing dog and does not reach maturity until about 3 years old, retaining his puppyish ways until this age and even beyond it. This can be very engaging at times, but totally inconvenient at others, and the shooting field is no place for it. So be prepared to make haste very slowly and while one may start gun training from 9 months onwards, one would not necessarily expect seriously to shoot over him until at least his second season. There is no doubt about the curly's intelligence and ability to learn, but experience has to be

17. A retrieve from a flight pond.

built up slowly. The other curly characteristic to be reckoned with in this context is that he has a mind of his own and is quite capable of using it his way. Intelligence and initiative are both admirable and necessary qualities in a retriever, provided they are channelled early in training the right way. The curly is obedient but tends to be so in his own time. Obviously to overcome this, the training of a curly calls for a very great deal of patience and anyone not of a very phlegmatic temperament would be well advised not to try. This independence of mind is one of the traits which curly devotees find attractive for it makes for a dog of character and most curlies are certainly that. Others find it less admirable to varying degrees; some have been known to call it pig-headedness. No doubt in the old days of seventy and eighty years ago the keepers whose dog the curly usually was had his own ways of coping with the problem. We must remember that at that time the word 'dog-training' was not used—the process was known as dog-breaking, which speaks for itself. Nowadays, dogs are not trained with the stick and the whip, but with psychology and the curly is as good a subject as any.

Early training in general obedience, before any gun training is contemplated, done with firmness and gentleness cannot be too strongly emphasized, for a curly allowed to go unchecked can very easily become wild and disobedient, a great nuisance and quite unsuitable for gun training. In view of the size to which the puppy will grow as an adult such a state of affairs would be intolerable. Unfortunately many people ruin a potential adult dog by being beguiled by a pup's natural charm into not disciplining it properly. Starting when he is a small puppy with generous amounts of praise when he does something right and a firm severe 'No' when he is doing wrong, he will learn what one respects and respect one's authority. The 'No' can be backed up when the occasion demands by picking up the pup a little off the ground by the scruff of the neck, shaking it sharply and growling. This is not cruel, but a natural way of disciplining it for it is what its mother would do. Later with a growing dog it is less easy physically to administer this kind of punishment, unless one has the build and strength. By that time, no doubt the pup has learned from previous experience and it is equally effective to scruff it and growl without actually lifting it bodily off the ground.

In training any dog, close study of the particular animal is essential if there is to be the understanding between trainer and pupil necessary for progress. It is essential, to quote a leading trainer of the present time, to get inside the dog's mind and to think like a dog and thus in training him to be always one jump ahead of him in whatever he may try to do. One therefore has to think like a curly.

The aim of training a retriever is to produce a dog that will be steady to shot and to dead and wounded game falling all round it, during a shoot, when it will walk quietly to heel and sit patiently, silently and watchfully until given the signal to retrieve. A dog thus trained is known as a non-slip retriever. This high standard of work is by no means easy to attain especially with a keen dog and if one feels it beyond one's capabilities as a trainer, then one should hand one's dog over to a professional trainer and should do so early. One must resist the temptation to start the dog off oneself for it is all too easy to start a dog on what prooves to be bad habits, which are far harder for him to unlearn, than new lessons are for him to learn. Give him basic obedience training and that is all; above all do not allow him to know what game, particularly live game is.

Having decided to have one's dog trained professionally, not any gundog trainer will do, good though he may be. It must be someone with knowledge, experience and above all understanding of curlies, and preferably with a liking for them and appreciation of them as working dogs. There are not many such trainers at present, but they do exist and are worth searching out. When one has been found, the dog is sent off to school and one must be prepared not to see him again for the best part of a year. A report every month or so will tell one how he is progressing and one must be content with that until the day one goes to collect him. Then, if all has gone well, one will find that the keen ignorant dog one sent away has been transformed into a smart disciplined worker who knows what he is about, as the trainer will demonstrate. This demonstration is important for it shows not only what the dog has learned, but also how he should be handled.

At this point one must remember that this is not the end of the business, but only the beginning. One now has a trained gundog, but as yet inexperienced, and no matter how well trained a dog is, he can be ruined in a few weeks by a foolish handler. So if one is an inexperienced handler, while the dog is away, one should have been doing one's homework by reading the best books one can find on retriever work. As a supplement a thorough discussion with the trainer when one collects the dog, on how he progressed, what to watch out for etc., will provide guidance on how to deal with him oneself when taking him out on the job.

A curly can be used for any kind of retrieving work, with any kind of game, as in the past, but it is in water work that he really excels. It is not necessary to have to teach him to get into water and to swim. As a small puppy he will walk and play about in any water he finds, thoroughly enjoying it. Later on one will find him swimming for as soon as he is taken near any sizeable piece of water he will be in, and swimming quite naturally, and soon he will become a really powerful swimmer.

The question of where to keep a gundog will always arise if he is to live with a family. It is not necessary for a working gundog to be kept at all times in a kennel, though it is useful to be able to shut him away there for mutual peace and quiet occasionally. Living indoors as one of the family does not impair a dog's working ability, and indeed he gains a great deal from being humanised from an early age. A curly is a sociable dog and enjoys human company and likes to be literally in contact with his owner, lying with a paw or his head on one's foot. My own often lies down and puts his nose on my feet while I am working in the kitchen or watching television. He gets on well with our spaniels. Usually dogs are happier living in company with other dogs, but they do not become humanised as much as when they are entirely dependent on human company. Our old Labrador who spent the early part of his life without other canine company, knew a great many more words and understood a great many more tones of voice than any of our subsequent dogs who have always had other canine companions.

In choosing a pup one would pick one that was bold, showing signs of self confidence and curiosity tempered with a degree of intelligent caution. It would be strong, well-boned, steady and balanced in its movements bearing in mind its age and allowing for a certain amount of puppy clumsiness. It would have to be capable of giving one its attention, if only momentarily at an early age, as puppies are easily distracted. A dog that will give one its eye as a young puppy suggests that he will continue to do it thereafter, an absolute necessity when training begins in due course. Rapport between dog and handler is essential for mutually successful training and subsequent working of a dog, and one that manifests it strongly at an early age is off to a good start.

The puppy would naturally be given the best of care and feeding to ensure healthy growth into the big strong dog it should be. A balanced diet of meal and raw mince plus vitamins, along with the dried milk specially made for puppies will ensure steady development. Starting at the rate of 4 meals per day at 8 weeks and gradually adjusting the size and frequency of the meals, he will as a young adult dog be having two meals a day; a main meal in the evening and a small breakfast. In these days of convenience foods there is no reason why a gundog should not be fed on them. They are formulated and balanced to contain all that a healthy dog requires. They come in pelleted form, can be fed dry and are very convenient to use and store. Choose one with a high protein content, and introduce it gradually into the diet, according to the manufacturers' instructions. It is obtainable in puppy size pellets so that puppies can be fed on it from weaning. With this type of feeding it is essential to provide the dog with free access to plenty of drinking water, particularly after a meal.

An outdoor kennel for a curly should have a sleeping compartment with a raised bed and a door to enclose it for warmth. Straw or sawdust bedding is quite unsuitable as it would get into the coat and be impossible to remove from the curls. Bedding is unnecessary for a fit dog. Rugs and blankets are too warm and are simply chewed up so newspaper provides the best solution if bedding is to be used, being cheap, clean and easily swept up if shredded. The kennel should also have an outdoor run enclosed by a high strong fence. It should be at least 6 ft. high, for curlies are very good climbers and jumpers, and preferably roofed for maximum security. The roof on our kennel is 7 ft. high and our dog regularly bounces when in it, touching the ceiling quite easily each time.

Needless to say a curly needs exercise and plenty of it. As a pup he needs plenty of free running exercise to develop good lungs and muscles which he will need as a worker and to work off the abundance of energy he will have as he grows. Daily walks on a lead will become part of his routine and will keep dog and owner fit. One must be prepared as the owner of a working gundog always to exercise it adequately out of the shooting season, especially if the dog is not worked regularly throughout the year. Too often the dog that works the winter season only is allowed to get fat and out of condition from January to September and gives a poor account of himself the following season simply through lack of fitness and poor management.

A working curly needs little grooming and never needs a bath, particularly if he swims a lot, for this would only remove the natural oil from his coat. The coat of a healthy curly does not have an unpleasant smell and it does not require trimming, as it does not grow to a great length, nor is it constantly being cast, covering everything with hair as in some breeds. Attention to cleanliness of ears, removal of burrs and thorns from his coat, occasional powdering to keep him free of parasites and the usual innoculations with regular boosters against diseases liable to be picked up from sheep and other live stock will keep him healthy. His sense of fun has been mentioned, so play with him but do not let him get wild and out of control.

Remember he is large, heavy and can be clumsy. He enjoys a romp and a game with a rubber ring or bone. Watch out for his tail, which can be like a whip lash if it catches one across the thighs and can clear a coffee or bar table with great ease and at considerable expense. In these days of female equality and women's lib, it is galling to have to admit that the curly is an old-fashioned dog and automatically respects a man, so that the man of the family has no trouble in establishing his authority. The woman of the household has to work for her authority because the curly will treat her rather differently and is not deceived by her trouser suit. He is quite prepared to be a tease, and she has to be prepared to dominate him by sheer personality—bullying would be useless—quietly confident in the knowledge that she will, with patience, be allowed to win in the end, backed up with a firm voice and manner, and stentorian tones when the occasion demands. Let the same woman, however, if she works the dog, put on a game bag and the other trappings of shooting and her curly will at once remember his training, that business is business and paramount. Together, handler and curly can be a useful addition to the shooting party.

If a large, handsome, intelligent, trainable, workable dog of independent turn of mind, with energy and stamina, both loyal and companionable, is what one wants, then a curly is a challenging and satisfying dog to own.

5 English Springer Spaniel

Of all our gundog breeds, the English Springer Spaniel is pre-eminent, both for its versatility and adaptability. A veritable 'maid-of-all-work', particularly suitable for the one-dog man who is an all-round shooter and seeks 'trainability', coupled with efficiency, docility and hardiness from his four-footed companion. Having bred, trained and shot over English Springers for more than 30 years, I have no hesitation in making this bold statement, but to it I must add one proviso—such paragons will only be obtained from *the genuine working and field trial strains* of the breed, a point which cannot be emphasised too strongly, and one which also applies when selecting a dog for work with the gun from *any* of the gundog breeds. Fortunate, indeed, that the working-type English Springer has proved its own point and become so widely popular that this presents little difficulty, whilst there are private and professional kennels throughout the British Isles which can supply good stock.

Unfortunately, like most of our gundog breeds, English Springers have, over the years, suffered from the attentions of the show bench fraternity who, being interested only in conformation and 'beauty', have tended to breed dogs for appearance rather than utility in the field, almost breeding out, in many strains at least, some of their most valuable natural working attributes. It is entirely thanks to a small but dedicated band of practical shooting men, who understood and appreciated the finer points of spaniel work and went to great lengths to retain them by selective breeding over many years, that today we have our splendid working strains. They put first things first and mated good workers to good workers, sometimes regardless of appearance and conformation, whilst the show bench folk did the exact opposite, ignoring the working ability of their breeding stock so long as it looked right.

Little wonder, therefore, that within the breed we now have two distinct types—show and working—and, it seems, never the twain shall meet! The show type is large, heavily-boned, narrow-headed and with long 'feather', deep flews and excessive eye haws. It moves ponderously and is often a reluctant, or at best a sloppy, retriever, with no great enthusiasm to face thick and thorny covert or water. The working type is almost the exact opposite—smaller, neater, more lightly-boned, with shorter ears and the minimum of feather, broad head with sensible lips for retrieving, and clean cut eyelids. It has a fast and stylish action and delights in thrashing covert and retrieving from land or water.

To be absolutely fair, there *are* individual show-type dogs which will put up an adequate performance in the field, and also working-type springers which might receive some recognition from a show judge, but these are few and far between! It is a sad but true fact of canine life if one tries to combine the two by breeding between them, you almost invariably end up with pups which exhibit the worst features of both types, as many enthusiastic breeders, with the best possible motives, have discovered to their cost.

It must not be inferred that our working English Springer Spaniels are ugly dogs—quite the reverse, in fact, as I hope the accompanying photographs illustrate. They are compact, athletic and of a sensible size for practical work, both questing in covert and retrieving, without the show exaggerations mentioned above. They bear far more resemblance to their progenitors of the distant past (which, remember, were sporting dogs pure and simple) than does the modern show-type animal. Apart from the purely physical differences between the two types (like comparing a cart-horse with a race-horse, some say!) the worker wins hands down on two scores—trainability and style, with which can be coupled pace and

18. The alertness of a keen Spaniel.

drive, particularly in difficult covert. In short, they are built for the job, physically and mentally.

The history of our spaniels seems to be shrouded in a certain amount of mystery, and even after a great deal of research into the subject I find it difficult to give chapter and verse of the evolution of the English Springer. The name Spaniel almost undoubtedly comes from Spain, so it is assumed that the dogs originated from that country, and some authorities believe that they were imported by the Romans when they invaded Britain, where they were used for flushing (or 'springing') game for hawking. There seems little doubt that spaniels and setters, originally, sprang from the same stock, and that the various spaniel breeds as we know them today were evolved by selective breeding and a certain amount of out-crossing.

In the year 1570, Dr. Caius wrote about spaniels in his 'Historie of Englishe Dogges' and described 'brown and black-pied Spaniells', and also introduced the term 'springer', indicating that these dogs 'sprang' their game for the hunter. The late Dorothy Morland-Hooper, in her book 'The Springer Spaniel', quotes a 17th century sportsman, Richard Surflet, as writing: 'The spaniel is gentle, loving and courteous to man more than any other dog, of free untiring laborsome ranging, beating a full course over and over, which he does with a wanton playing taile and a busie laboring noise, neither nor showing less delight in his labours at night than he did in the morning'. This, with the exception of the reference to 'a busie laboring noise' (spaniels are now expected to hunt mute) aptly describes temperament and the working ability of the modern working English Springer.

The development of the shotgun, undoubtedly, increased the value of spaniels to sportsmen, and various breeders, with differing ideas and requirements, started to produce dogs which were suitable for their particular form of shooting, and the country over which they shot, and thus were evolved the several spaniel breeds as we know them today. Many of these, unfortunately, have, of recent years, taken very much of a back seat in sporting circles, again thanks largely to the influence of the show bench people and because the dogs got into wrong hands. What we now know as the English Springer was more fortunate, and Dorothy Morland-Hooper gives in her book what I believe to be the most authentic account of its early development, which I quote hereunder.

'About 1800, the Boughey family of Aqualate, in Shropshire, developed a distinct strain of Spaniel and bred it very carefully, keeping a stud book from 1813 onwards for the 'Springer Spaniels'. For over a century the strain was kept by successive generations of the family, and in 1903 Sir Thomas Boughey bred F.T. Ch. Velox Powder, later owned by Mr Eversfield and winner of twenty F.T. stakes, whose pedigree goes right back through the Aqualate Stud Book to Mop 1, whelped in 1812 . . . The Boughey family continued its interest in English Springers right into the 1930s and I remember Mrs Boughey, wife of Commander E.V.F.

Boughey, telling me that Sir Francis Boughey once used a French Poodle in his breeding to secure extra sagacity . . .'

Miss Hooper also makes reference to the fact that during the latter part of the 19th century there were quite a number of typical Springers described as 'Norfolk Spaniels', 'perhaps because such a large number of them were in Norfolk and spread over the country from Norfolk'. We also find mention of the Norfolk Spaniels in 'The Sporting Spaniel', by C. A. Phillips and Claude Caine, first published in 1906. 'To many of our readers the Springer will be more familiarly known by the name of Norfolk Spaniel, and under this name it was given a separate classification and scale of points by the Spaniel Club. It was, however, a misnomer, and Mr Farrow rightly pointed out that the breed had never been confirmed to one county, like the Sussex Spaniel, and that its only connection with the name of Norfolk arose from the fact that a former Duke of Norfolk had possessed and fostered a notable strain of this variety of Spaniel. The breed was accordingly deleted entirely from the Spaniel Club's classification, though in the opinion of many members this was a mistake, as all that was, necessary in their opinion, was to change the designation 'Norfolk' to that of 'Springer'.

To complicate the issue even further, Mr Phillips himself, around about 1910, mated a Springer dog, Spot of Hagley, to his Cocker bitch called Rivington Riband, and a dog from this litter, Rivington Sam, became the first English Springer field trial champion in 1913! Subsequently Sam was widely used as a sire, and, like the aforementioned Velox Powder, can be regarded as a pillar of the breed, to whom many of our illustrious working dogs of the present day can be traced back. The Kennel Club first recognised Springers as a separate breed in the year 1902 and, apparently, did not regard this infusion of Cocker blood as anything out of the ordinary, as would be the case today. Be that as it may, it was a very good thing, and we should all be grateful to Mr Phillips, along with a host of other early spaniel breeders, far too numerous to mention, for their foresight, dedication and selective breeding programmes which combined to 'manufacture' a utility dog of supreme value to the shooting man, probably more appreciated today than ever before. They laid the foundations of the breed very, very well indeed.

The institution of field trials in 1900 gave great impetus to the English Springer Spaniel which, over the next decade, emerged as a serious contender for top honours. This encouraged more shooting men to take up the breed, and inspired owners to examine and improve their breeding policies. At this period, most of the top breeders also had an interest in showing their dogs on the bench, but never seemed to lose sight of the fact that they were *gundogs* and, as such, should combine brains with beauty. This is something which, alas, our present-day show bench fraternity seems completely to have forgotten!

It is difficult to put a finger on an exact date when the rot set in, and the 'show fancy' started to breed for the (to me) repulsive exaggerations of type which the *show* springer of today exhibits, but the best guess I can make is 1921, when the English Springer Spaniel Club was formed after a meeting held at Cruft's Show, although perhaps the majority of supporters at that time were enthusiastic shooting men and women. This Club did much to popularise the breed with 'ordinary' owners—perhaps *too* much! and English Springers were taken up by pet and show folk who had no interest in, or access to facilities for, training their dogs to the gun. Thank heaven for those stalwart pioneers who continued to put first things first and who, against massive opposition, succeeded in retaining —and also improving—the essential practical qualities of their dogs, even if it did result in the emergence of a distinct 'working type' within the breed.

It might be interesting, here, to quote from the Kennel Club standard of points for the breed as published by the English Springer Spaniel Club. This states: '*Characteristics:*— The English Springer is the oldest of our Sporting Gundogs and the taproot from which all our sporting land spaniels (Clumbers excepted) have been evolved. It was originally used for the purpose of finding and springing game for the net, falcon or greyhound, but at the present time it is used entirely to find, flush and retrieve game for the gun. The breed is of ancient and pure origin, and should be kept as such. The Springer's gait is strictly his own. His forelegs should swing straight forward from the shoulder, throwing the feet well forward in an easy and free manner, not a paddle nor a chopped terrier-like stride. His hocks should drive well under the body, following in a line with his fore-legs. At a slow movement many Springers have a pacing stride typical of the breed.'

19. Three famous names in English Springers. (Left to right) the late John Forbes (Glennewton, John Kent (Chrishall), and Joe Greatorex (O'Vara).

5. English Springer Spaniel.
John Marchington

6. English Cocker Spaniel.
John Marchington

7. Irish Water Spaniel.

8. Welsh Springer Spaniels.
Peter Moxon

9. Clumber Spaniel, Sh. Ch. Raycroft Skipper.
Miss A. Roslin-Williams

It is quite impossible to include a list of *all* the top kennels, owners, handlers and dogs, past or present, to whom we are so indebted, but mention must be made of such early kennel prefixes as 'Avendale' (owned by the late Duke of Hamilton), 'Rivington' (Mr C. A. Phillips), 'Laverstoke' (Lady Portal), 'Denne' (Mr C. C. Eversfield), 'Tedwyns' (Mr A. Byrne), 'Banchory' (Countess Howe), 'Cairnies' (Mr D. L. Wigan), 'Beechgrove' (Mr Winton Smith), many of which will be found, if traced back far enough, in the pedigrees of our best present-day working English Springers. So, too, will 'Blair' (Mr G. Clark), 'Bryngarw' (Capt. Traherne), 'Staindrop' (Edgar Winter), 'Wittlesford' (Col. C. McNeill), 'Wakefares' (Mr Mason Prime) and 'Pierpoint' (Mr W. D. Edwards).

Without any shadow of doubt, the greatest single influence upon working English Springers of more recent times have been the illustrious O'Vara dogs, owned by the late Mr Selwyn Jones and handled by Mr T. J. (Joe) Greatorex. Both Mr Jones and Joe had an uncanny knack of picking the right dogs for work (and, equally, of culling the unsuitable ones), and their really sound breeding policy (with particular emphasis upon the female line) quickly established the O'Vara dogs as *the* outstanding strain between 1928 and 1953. Their record of field trial wins, both in this country and abroad, was phenomenal, and right up to the present time breeders of working springers boast of the number of O'Vara dogs appearing in the pedigrees of their stock.

F.T. Champions Don and Spy O'Vara are names which crop up again and again in working pedigrees, litter brothers from the 'foundation union' of a dog called Menaifron Jock with a bitch named Flicker. Joe Greatorex, now in retirement in Berkshire, was a truly great trainer and handler, who always succeeded in making his dogs 'part of him' and, 'thinking like a dog', managed to keep one jump ahead of them. This psychological approach undoubtedly accounted for much of his outstanding success.

Other breeders and trainers who have done much for English Springers in recent years include John Kent, whose Chrishall dogs made great impact both in this country and abroad; the late John Forbes, of Glennewton fame (and laterly handling the Elan dogs of Commander Collard); the late Hal Jackson, a very 'professional amateur', who made his Barnacre springers a household word; the late Bill Sheldon whose Ludlovian dogs were such natural workers; that irrepressible Welshman, the late Tom Evans, of the Garwgarreg prefix; the late Jack Curtis (Whittlemoor); also Mrs P. M. Badenach-Nicholson (Carswell); Tom Laird (Criffel); Talbot Radcliffe (Saightons); George Curle (Breckonhill); Jack and Keith Chudley (Harpersbrook); Dr. Tom Davidson (Jonkit); Keith Erlandson (Gwibernant); Mr W. Charles Williams and his handler Dick Male (Berrystead); John and Peggy Pratt (Brickclose); Bernard Dutton (Hamers); Andrew and Jim Wylie (Pinehawk); Jack Davey (Wivenwood); Dr Diana Bovill (Harwes); Danny MacKenzie (Staxigoe); Esther Thompson (Rivington); Frank Bell (Adee); Jack Davis (Gowna); Major Peacock

(Greatford); and Arthur Cooke (Hales). (It is, perhaps, worthy of note that most, if not all, the dogs bearing the above-mentioned prefixes will be found to strain back to the O'Vara line, if pedigrees are closely studied!) These are all names for which to look out for in pedigrees of working dogs.

British working and field trial Springers have, since Hitler's War, proved very popular exports, to shooting and trial exponents, all over the world, notably in the United States of America, Canada, Australia and Italy. So much so, indeed, that some authorities believe that we have seriously denuded our superior breeding stock through sheer commercialism, and that this will have a serious effect upon the breed in the near future. Whilst I agree that some of our best blood has already gone to other countries, I am convinced we still have enough in Great Britain to carry on the good work. What, today, we probably lack more than good dogs is *good trainers*, and also breeders of the calibre of past pioneers, willing to formulate a sensible policy, stick to it and cull the 'rubbish' ruthlessly, instead of selling it off cheap (or even giving it away) to people who know no better, so perpetuating serious mental and physical defects, which the pioneers of the past strove so valiantly to eradicate.

CHOICE AND SELECTION OF A PUPPY, STARTED OR TRAINED DOG

Having, I hope, 'sold' my reader upon the idea of acquiring an English Springer Spaniel as a shooting companion, a few words of advice upon picking the right animal may not come amiss. The first essential is that your dog, of whatever age, comes from a genuine working background, or you will be bitterly disappointed. So, obviously, you must contact a breeder or trainer of repute, who has everything to gain and nothing to lose by supplying you with an animal which will fulfill your purpose and do him credit. Unfortunately, dogs are often advertised as being 'of working strain' when, in actual fact, they are nothing of the sort.

If you have no contacts among established breeders of working-type springers, or a knowledgeable friend who can be relied upon to give you good advice, then I suggest you write to The Secretary, The Kennel Club, 1 Clarges Street, London W1Y 8AB, and ask him to forward a list of breeders of field trial dogs. Other useful contacts would be the Hon. Secretaries of both The Spaniel Club and The English Springer Spaniel Club, c/o the Secretary of The Kennel Club, either of whom will be pleased to put you in touch with likely suppliers. Do not be put off by my mention of 'field trial' breeders because you have no intention of using your dog in competition. The point is *that you want the best possible dog for work with the gun*, and you cannot go far wrong if you obtain one of impeccable, award-winning background.

Choosing a puppy from a litter at between eight and twelve weeks of age, when most are usually offered for sale, is not an easy matter, even for

20. The working-type English Springer Spaniel has shorter, less 'feathery', ears than the show-type.

the real expert, and I always remember Joe Greatorex, many years ago, advising me that in order to be sure of picking the best for work you have just got to run on the whole litter and start the training, making the final selection at nine or ten months of age. This may be all very well for the breeder and trainer, but quite impossible for the ordinary shooting man. So you have just got to be guided by commonsense and instinct or, more probably, by your wife's intuition! This is something which, incidentally, I have discovered over the years to be extraordinarily accurate in this direction.

Imagine that we are looking at a litter of six to nine puppies, and have been offered the choice of them. No doubt you will already have decided upon which sex you want, and also the colour, either liver and white or black and white. You may also have some predisposition as regards size and shape—you may require more white than colour or *vice versa* in your puppy. These considerations, obviously, narrow your choice, so you go on from here. I suggest that you see all the puppies running about in the yard, or on the lawn, both with and without their dam being present. Watch how they move, note the friendly and the shy ones, clap your hands loudly to see if any appear noise-shy and, above all, try to spot the one which, for some possibly indefinable reason, attracts you. Then, having got down to a 'short list', ask the breeder to demonstrate their natural retrieving ability by throwing a small dummy for them to fetch, one by one, and make your final decision accordingly.

THE 'STARTED' YOUNGSTER

What are the prospects of finding an older puppy—one which has had the rough edges filed off, is over the inevitable 'teething troubles' experienced with a real youngster, and, possibly, one which has learned at least the rudiments of good behaviour and gundog work, and can therefore be put into serious training right away? Here the owner must tread very warily indeed, because unless dealing with a really reliable source, he may be very disappointed. To put it bluntly, the average breeder-trainer prefers to sell his young stock as soon as weaned, only retaining those puppies which he believes to be likely prospects to run on and train, either for his own use or for sale as fully-made dogs. Along the way he weeds out the 'second-raters', and usually offers them for sale at appropriate prices, according to their aptitude for and performance in training. So, whilst it is often possible to obtain a very adequate *shooting* prospect, perhaps being sold because it does not show those extra qualities required for field trials, aged between six and eighteen months, you may equally well end up with a very poor specimen, unless the seller is a reputable man with a name to lose.

My suggestion is that, having found a likely source of supply, you go along and see the dog or dogs offered. Discuss them with the owner, and ask him to put them through their paces, demonstrating all those abilities which he claims for them such as questing, simple retrieving, sitting to command and/or whistle, reaction to gunfire. If you are very inexperienced, then I suggest that you take along a more knowledgeable friend to 'vet' the dog for you, and give his frank opinion.

If you decide to buy, make sure that you obtain a *written* list of commands, signals and whistles to which the dog is accustomed, and also a whistle of the correct type. Few, if any, reputable breeders and trainers will allow their dogs to go out on approval, because of the danger of them being ruined by mishandling. All the seller can do is to demonstrate how the dog will work for *him*, and it is then up to you to make the dog feel at home, by tact and ingenuity, and to follow the seller's instructions to the letter. It may take days, possibly weeks, before the dog fully settles down and accepts you, so be patient and understanding, and do not attempt to do too much too soon.

THE FULLY-TRAINED DOG

The above recommendations apply with equal force if you are seeking a fully-trained dog, and you will, naturally, expect the vendor to give you a reasonably comprehensive demonstration, preferably using real game for questing, retrieving and steadiness. If possible, see the dog working in cover and water, and satisfy yourself, as far as is possible, that it has a tender mouth. Ideally, you will watch the dog working naturally on a shooting day, but this is seldom possible, and well-equipped trainer-breeders usually have excellent 'artificial' facilities, including a rabbit-pen, wherein the prospect can demonstrate its abilities. Always find out from the vendor any little peculiarities the dog may possess, watch him like a hawk, pick his brains and try to emulate his handling methods (certainly always using the same commands, whistles and signals), so that the change-over will go smoothly for both the dog and yourself.

The most common and serious mistake made by new owners of fully trained dogs is taking them out shooting much too soon, without giving them any chance to become accustomed to their strange handler and environment. I have known dozens of cases when a dog has been taken over one day and out to the shoot on the next, with the inevitable result that either the dog refused to work at all, or committed every crime in the book. Invariably, when this happens, the owner blames everybody but himself, mostly the trainer from whom he acquired the dog, with general dissatisfaction and unpleasantness all round. Many a dog has been ruined, for all time by such stupidity on the part of the owner and, at best, it may take weeks or even months of tactful handling to undo the harm which has been caused.

Any trainer who knows his stuff goes to great lengths to explain to his clients that a dog is not a machine, and that it cannot possibly be expected to adapt itself to entirely new surroundings and handler in less than a couple of weeks, or maybe considerably longer, no matter how well trained it may be, or how much was paid for it. There are occasional exceptions (the ones which prove the rule!), but these are the very stolid and adaptable dogs which get into the hands of men who have that rare commodity, 'dog sense', and are capable of immediately winning over the respect and affection of any dog.

There is only one answer, namely to spend time and trouble in making a strange dog completely 'at home', first of all in its new environment and then, stage by stage with the handler. My advice is that the dog is given a few days without any handling, other than exercise (at first on the lead), and then gently practised in the simple obedience and dummy retrieving exercises close to home—on the lawn, for instance. If all goes well, build up gradually from here, and when you are confident that the dog has accepted and 'taken to' you, and has transferred its affection from its former handler to yourself, extend the operation, taking it out with the gun, alone to start with or, better still, without carrying the gun yourself but with a friend to do the shooting, so that you can concentrate upon the dog, maintaining steadiness and discipline.

The next step is to take the dog to a shoot, again without yourself carrying a gun, in the capacity of beater or pick-up, or a combination of the two. Thus your entire concentration is centered upon your dog, which, believe me, will know this and behave accordingly. *When, and only when*, you are reasonably certain that you have 'got on top', and gained the confidence of the dog whilst gunless, you can start shooting over it (alone at first, preferably), and then try to teach yourself the difficult feat of keeping one eye on the dog and one eye on the game when you shoot!

Always be prepared to give the command to sit immediately game rises, and again on shot if neccessary, and *never, never send the dog to retrieve immediately a bird falls*, but give it a period of waiting on the drop first, even if the bird is a runner. Neglect this and you will have an unsteady dog on your hands in less time than it takes to tell.

HAVING A DOG TRAINED

Some shooting men lack time or confidence to train their own dogs, and do not fancy the idea of buying a trained animal. Instead, they opt to obtain a puppy of good working blood and send it away to a professional training establishment, when it is old enough, to be educated according to their own requirements. These days professional public gundog trainers are very busy men indeed—there are really not enough good ones to meet the demand for training. They usually require advance bookings for pupils, so it is advisable to get your puppy's name down at a good 'school' as soon as you acquire it.

Most trainers like to have the dogs at around six to nine months of age, and a thorough intensive training course is likely to last from four to six months, according to the ability of the pupil, the standard of training desired, the weather and . . . luck!

Having made contact with the trainer of your choice, you should ask him if he prefers you to give your puppy any basic education before going into kennels. Some men encourage their clients to make a start with simple obedience and dummy retrieving, others prefer completely 'virgin material'. What you must certainly *not* do is let the puppy get into bad habits, go off self-hunting, or make it too 'soft'. It is helpful if the animal has become accustomed to living in kennel—even if only for short periods—and it must be fully vaccinated against the major canine diseases.

During my training career, I suffered enormously through well-intentioned clients trying to save themselves money by 'making a start' on

21. Working-type English Springers are excellent retrievers from water.

their youngsters before sending them to me. The result is that I have had pupils come in with hard mouths, gun-nervous, gun-shy, or so completely wild (having already learned the delight of chase and that the 'bang' of a gun means something falling for them to retrieve) that they were difficult, if not completely impossible, to educate to anything like a reasonable standard. I have suffered, too, through the dogs being of entirely the wrong breeding for work, and most sensible professional trainers insist upon scrutinising the pedigrees of their prospective pupils before accepting them for education.

I have also suffered from owners 'breathing down my neck', and constantly prodding me (and coming to visit their dogs) during the training course. So select a trainer of repute, take the dog down to him and then leave well alone until you are invited to go along for a 'half-term visit'. Remember that, in gundog education, progress is not, and cannot be, smooth and continuous, and that no established trainer will keep your dog for a moment longer than necessary due to the pressure of his waiting list. He will obviously, for the sake of his jealously-guarded reputation, do everything possible to make a success of your puppy. No trainers like failures, and all the good ones are conscientious to a fault. Most will accept dogs for a probationary period of about four weeks (two of which, at least, will be spent in the 'settling down' process), after which you are advised if you have a likely prospect or not. A dog may be rejected (for an incurable fault) at any time during the course.

Assuming that all goes well, and that your English Springer puppy trains on according to hopes, you will be advised by your trainer when it is ready for collection. You will if you have any sense at all go down to the kennels

22. English Springer well camouflaged on a snowy marsh.

23. A rabbit well retrieved.

to pick it up, and learn from the trainer all that you should know about the dog. Assuming that you acquired the puppy at an early age, and kept it at home for some weeks, or maybe months, before sending it away, it will not have forgotten you, and when you get it home and it has had time to sniff around and recall all the old smells and sights, the settling-in period should not prove difficult.

However, having written this, I emphasise that you should *proceed in accordance with my suggestion above for dealing with a purchased fully-trained dog*. However well the dog appears to remember you and the family and it will almost certainly try to take advantage of the family's affection for it, just like a child back from boarding school, do not let it start handling you! Be patient, yet firm and, above all, *just*, and do not make the mistake of taking it out into the shooting field until you have completely transferred its obedience and respect from the trainer to yourself. If you run into any problems, do not be too hasty, but contact the trainer and ask his advice, remembering that it is in his interest to ensure that his efforts over many months have not been wasted.

In this way, you stand every chance of making a complete success of your dog but, please do not let your wife and/or children spoil it by over-indulgence. There is no earthly reason why your dog should not combine the dual role of practical gundog and family companion, so long as your general management is sensible and you train other members of the household not to interfere when you are handling it. They must all realise that, if a command is given, it should be obeyed. Above all, you are the dog's lord and master, and, as such, must remain its main centre of interest.

FEEDING AND GENERAL MANAGEMENT

A dog's natural diet is, basically, raw meat. However, these days raw meat is both difficult to obtain and expensive, and, in any case, you may have preconceived ideas about the rations you are going to feed your English Springer. Suffice, on this subject, to say that you must be guided by commonsense, the depth of your purse and the advice of the breeder and/or the trainer of your dog. I prefer to feed raw meat including tripe and other offal to my dogs whenever possible, but in recent years, I have been forced, for practical and economic reasons, to change over to one of the modern, 'all-in' dry, pelleted dog foods, which is not only clean and convenient to feed, but supplies all the nutritional requirements of the dogs and 'does' them very well. I have now used 'Evenlode Dog Food' (manufactured by Messrs. Evenlode Products Ltd., Evenlode, Moreton-in-Marsh, Gloucestershire) for over two years, and have found both the adult and puppy rations to be everything that are claimed for them, being very palatable to the dogs and maintaining them in peak condition.

Generally speaking, English Springers are very good feeders, even on the greedy side, and this form of diet suits them particularly well. I believe in two meals per day, even for adult dogs, rather than the traditional single feed, especially when in hard work.

General management of your Spaniel depends upon your circumstances and environment, but certain basic principles should be borne in mind. Dogs are creatures of habit and respond best if a regular routine is imposed. English Springers are particularly hardy and adaptable, in the main 'sensitive but sensible', thus making ideal household companions as well as workers. They will do well both living in kennels or indoors, and I heartily recommend a combination of the two systems as being best for both the dog and the owners.

If, from an early age, a puppy is accustomed to spending at least part of the day in a properly constructed kennel and run, free from draughts, dry and with adequate exercise facilities, it will do well and never prove to be a problem at those times when it is essential for it to be out of the way of the family. You can spend a fortune on building a kennel and run, or you can adopt part of the garage, or other outbuilding, or purchase a custom-made, sectional kennel, with or without exercise run. If possible, the run should be long and narrow, rather than square.

Small and snug sleeping quarters are the best answer, in my opinion, and I do not recommend straw or other bedding. This only encourages skin vermin and, if allowed to get damp, causes all sorts of troubles. I have kept my dogs, for many years, on bare boards and they have done very well under this system. I use dry sawdust which is changed every day, on the kennel floor, and, of course, the exercise runs are cleaned out daily.

Grass exercise paddocks are ideal for puppies, so long as they are well drained and the perimeter wire is dug, or concreted in to prevent burrowing and is sufficiently tough to withstand chewing. Concrete or brick runs are more convenient and hygienic for adult dogs, and escapism must be guarded against in both kennels and yards. Wire must be tough, tall and, where possible, provided with a 'turn-in', and chewable sections of the kennel protected with galvanised metal sheeting.

Exercise periods must be regular, like mealtimes. English Springers are very active dogs and do best when given plenty of free-running exercise, especially first thing in the morning and last thing at night. Gundogs in particular enjoy their freedom, and, in common with all dogs, probably think more highly of the person who gives them this than they do of the one who feeds them. For this reason I particularly recommend that whoever is going to work the dogs should also be responsible for giving them free-running exercise, and this applies both to puppies and adults.

Every dog (as a puppy), should be vaccinated against the major canine ailments, as recommended by your veterinary surgeon, and also wormed from time to time, and your whole treatment of it should be consistent and just. Things like brushing and grooming, bathing and dusting with insect powder must depend upon circumstances and conditions, but should not be over-done, whilst general hygiene and disinfection of kennel quarters are obvious routine matters. Successful dog-keeping depends largely upon the commonsense of the owners, who must remember that dogs cannot *tell* them what their requirements are. Try always to keep 'one jump ahead', and to 'think like a dog'!

TRAINING

The English Springer is unquestionably the easiest spaniel to train for work with the gun, and the most efficient when trained, *if of genuine working strain.* It is, obviously, impossible to go deeply into this subject within the confines of a single chapter, but I will try to put readers into the picture in as few words as possible. There are plenty of good text books on the subject available, so that anyone who becomes bitten by the training bug can read it up, and stands every chance of making a success of his puppy. My own book, 'GUNDOGS: Training & Field Trials' (Popular Dogs Ltd.), now in its tenth edition, is unblushingly recommended, and deals exhaustively with the whole process.

What is required of the English springer in the field? Normally the breed is best utilised for questing unshot game, flushing it and retrieving the slain and wounded. For this purpose the dog must hunt within gun-shot range, remain steady to flush, shot and fall, and go out to retrieve only upon command. Springers *can* be trained, and very often are, to act as no-slip retrievers, both walking-up and at drives, but, to be frank, this is not really their scene. Certain individuals take kindly to this sort of work, and perform it well *if given specialised education early on in life*, but the great majority are, by nature, restless and energetic, and want to get on with their proper job of questing, having, for generations, been bred for this specific purpose, at which they undoubtedly excel. So, if you require your dog as an 'all-rounder' and to be quiet and steady on those occasions when you are shooting at drives, walking up or operating from a duck or pigeon hide, you must bear this in mind from the commence-ment of training and educate it accordingly.

The king-pin of training is *obedience*, and you have got to channel a dog's natural instincts to serve your purpose. There is a great deal of psychology in gundog education, and the puppy must be taught right before it has had the opportunity to learn wrong. The trainer should be the centre of attraction, from the very start, and his word has got to be law, brought about in such a way that the pupils' natural instincts are not inhibited. The art of training is in knowing just how far to go and when to stop, and in learning to understand the individual pupil. Keep the puppy happy, making the lessons enjoyable and as much like a game as possible. Short—10 to 15 minutes—lessons, frequently given, are better than long sessions. Boredom must be avoided at all costs, because it causes a dog to lose enthusiasm, style and drive, making it stale and unreceptive. This applies particularly to sitting and retrieving lessons. Over-use of the dummy on simple, in-sight retrieves is a common and dangerous mistake made by many novice handlers. Retrieving especially should be a privilege, not a penance, for the pupil.

The training course can be roughly divided into two sections—preliminary and advanced, and experienced trainers are agreed that *the preliminary course is by far the most important*. Here you are dealing with a dog in its formative months and laying the foundation for all future behaviour. It is essential that you get on terms with your puppy, gain its affection and, more important still, its *respect*, learn the priorities and get them right. Then, by a progression of exercises, educate it in its future duties in such a way that it does not realise that it is being trained!

The age at which to commence serious, intensive training must depend upon the temperament of the particular puppy. Generally speaking, bitches mature more quickly than dogs, and can often be started earlier. Bold, forceful puppies, of either sex, can be ready for mild discipline at around five or six months. Shy and sensitive ones (often the most efficient mature workers) need more time (and tactful treatment) and should be allowed to reach eight, nine or even ten months of age before putting the pressure on them. See what I mean by getting to know your dog?

However, for all this, much spade work can be done at a very early age—almost from the nest stage, in fact. Coming to call and whistle, retrieving a small dummy (a knotted handkerchief or piece of rag is useful here), and accustoming to light cover—long grass, for instance—can all be practised during the short periods of exercise required by a young puppy. What you must *not* do is to risk making your puppy 'sticky' by giving it sit and stay lessons before it is ready for them. By all means accustom the youngster to travel by car, to walk on a lead (preferably a cord, 'choke' type slip lead), and to meet as many people and situations as possible. In other words, humanise it and build up confidence, which will prove invaluable when it is first taken out shooting in company.

I am a great believer in making a puppy as much 'part of you' as possible during these early, formative months, but, equally, I am convinced that, *once serious and intensive training commences*, the pupil will do much better if kept in kennel, and not in the house, so that it comes out fresh—mentally and physically—when training-cum-exercise is on the programme. In this way the dog looks forward to its education, and also thinks all the more of its handler and is receptive to the lessons that he gives, whereas one which has been hanging around the house and garden all day is liable to be stale and disinterested. This is another piece of 'canine psychology' which, if you think about it, makes real sense! This is why I recommend that, from the start, every gundog should become accustomed to kennel life. English Springers are one of our hardiest of breeds, and thrive under quite Spartan kennel conditions if properly managed. Apart from this, it is so convenient to have a dog which is equally at home in the house or in its kennel. Furthermore, a dog which has free range of house and garden at all times is very liable to become bored, and to wander off and get into all sorts of trouble—from scrounging local dustbins to self-hunting or sheep worrying–inimical both to training and good neighbourly relations!

I make no apologies for this long preamble—call it a sermon if you like! —to training. Experience has taught me that so many folk make serious mistakes with their first dog, or dogs, simply because they are in too much of a hurry and lack forethought in their whole approach to the subject. Training is easier than you think, so long as you try to see things from the dog's point of view, read the subject up, exercise patience and tact, and remember that dogs have limited intelligence and cannot reason as we do.

The preliminary course consists of encouraging the puppy to retrieve a dummy to hand, teaching it to sit upon command, whistle and signal, to stay until told to move, and to quest in front of you, from side to side, within reasonable gun range (up to 25 yards or so), so that anything that it flushes will be within shot. All this can be achieved, if you follow your chosen textbook carefully, on the lawn or in a suitable field, under strictly controlled conditions and without live game being present. So, too, can *tactful* introduction to gunfire (using a blank-cartridge pistol and/or a 'dummy launcher' which provides a bang and lofts a dummy at the same time, and is particularly useful for teaching a puppy to mark), jumping fences and facing cover, if you are prepared to use a bit of ingenuity in providing such 'obstacles'. Work in water can be introduced into the curriculum when the weather is suitably warm, on a pond or stream where the banks shelve gently.

Once your dog has learned the basics of its job by artificial means, is thoroughly obedient and controllable, *but not before*, it can progress to the advanced training course, where it will be introduced to the real thing, hunting for and retrieving natural game in the field. The change-over must be very gradual and carefully controlled, and the handler must always try to keep one jump ahead of his pupils, and be prepared to 'put the clock back' should things go wrong. The change-over from artificial to natural training is a very critical period in a gundog's life, where novice handlers are prone to make mistakes, and their dogs to receive the wrong impression. So, once again, I implore readers to heed the advice of the book which they are following. All the good ones go to great lengths to emphasise this very point, and give clear instructions how to proceed. Of course, no author, however experienced, can give *all* the answers to *all* the problems which may be encountered, because so much depends upon the individual temperaments of both pupils and handlers, to say nothing of the circumstances which may arise during the training course. In training a gundog, one learns something new every day, and never knows it all. This makes the whole effort so very worthwhile and interesting.

The countless English Springer Spaniels which I have had through my hands for training over the last thirty years have taught me a lot, and I am still learning. As a public professional trainer, and a keen practical shooting man, I have been fortunate to have the opportunity to train and shoot over almost every known breed of gundog, and also to watch them in action at field trials all over the British Isles. This prolonged and enjoyable experience has taught me that, without any shadow of doubt, the *working type* English Springer Spaniel is the most versatile and efficient gundog for the average, all-round shooting man, and especially the rough-shooter, who likes to train his own puppy for work in the field and, at the same time, seeks a biddable personal and household companion, fulfilling the dual role of worker and pet.

6 Cocker Spaniel

Those who have read 'Jorrocks, or the Sporting Tomlinson' by S. F. Outwood will be well aware that the Old Gentleman who commands the nether regions, is a keen hunting man, hunting the bagged M.F.H. three days a week with a pack of mixed Hell Hounds. Fewer of us, I think, are aware that he is also fond of his shooting and when he indulges in this sport, shoots over a pack of Cockers, black Cockers, each one with two tails and live coals for eyes. Every one runs in to flush, shot and fall. Each one gives tongue and all bite their birds like fury. Once in a while, their owner sends his foulest stud dog to earth to mate a few selected bitches, in order that the qualities of his chosen breed may be perpetuated, to the consternation of man.

On a more factual note, it would appear that the progenitors of the Cocker are probably the oldest of the gundog breeds in Great Britain. A Welsh king, Hywel Dda (Howell the Good) who ruled in the tenth century A.D., unlike most of his contempories, appears to have been more preoccupied with the welfare and happiness of his subjects, than with quarreling with his neighbours. King Hywel is mainly remembered as a great lawgiver, some of these laws relating to Spaniels, though in what context I cannot be certain. However, one fact of which I am certain is that the king loved hawking and his falconer was fourth man in the hierarchy of Wales; upon the entry of the royal falconer into the banqueting hall, provided the day's sport had been good, the king would stand and greet his subject. Despite the falconer's somewhat exhalted position, we are told he was only allowed a maximum of three drinks, lest he should neglect his hawks. Let us hope they were large ones! When we consider the king's preoccupation with falconry and the fact that Spaniels were definitely in existence at this time, we can safely assume that Spaniels were used in conjunction with the falcons for flushing their quarry.

It has also been suggested that Spaniels have been used over the centuries in Wales for flushing game into strategically placed nets, so could it be that Cockers or Cocking Spaniels, became so named owing to their employment in flushing woodcock into ride nets? It seems certain that these old Welsh Spaniels are the ancestors of the modern Welsh Springer and the present day Cocker. Contemporary evidence suggests they were red and white, so it would appear that the Welsh Springer has bred more true, at least in colour, than has the Cocker, although these colours do still appear occasionally. But somehow, since those early days, the Cocker has appeared in several additional colours—liver and white, black and white, black, white and tan (or rust), blue roan, red roan, golden and solid black, not to mention the solid livers which are so difficult to see in tall bracken.

Originally, the Cocker was simply a working Spaniel under 25 pounds in weight and the breed did not receive official recognition until 1893. Apparently a famous black dog, Ch. Obo, made his appearance in 1880 and won very consistently on the show bench. He was used very extensively at stud and it is said that the best present day blacks are descended from him. One does not have to be too concerned that he was a bench dog. Spaniels were shown long before field trials for them came into being as, for an appreciable period, no two owners could agree upon what precisely were the duties of a working Spaniel. I would imagine that the purists simply expected a Spaniel to hunt, find and flush the game and drop to shot, his work finishing at this point and the game being retrieved by a Curly or Flat-coated Retriever. I have seen a highly improbable sketch entitled 'A Canine Whipper In', which depicts a recalcitrant young Cocker, having run in to fall, being retrieved by a huge, grim looking Curly-Coat at the command of the gamekeeper. The explanatory text quite seriously vouched for the authenticity of the story.

I imagine the more adventurous and less hide-bound sportsmen insisted that a Spaniel should retrieve to hand also. When, many years ago, I visited the famous Rivington Kennels for the first time, I saw pinned on the rafters, a great many show cards besides field trial awards. I asked my friend, the late Colin Thomson, son of the late James Thomson, trainer to the late Mr C. A. Phillips, what show cards were doing in a famous working kennel. He informed me that the Rivington dogs were formerly shown, before the emergence of field trials and that, in those far distant days, they simply showed their working dogs, who were bred for work first and foremost, the division of the gundog breeds into show and working types occurring many years later.

The blacks, descended from Ch. Obo, dominated the show ring for many years and it became the object of Mr C. A. Phillips to breed coloured specimens of quality equal to the blacks. He acquired two Welsh bitches from Mr James Freme, of Wepre Hall, Flintshire, one of which proved to be the founder of a successful coloured line. This blood spread to other successful kennels and probably accounts for the great diversity in colour encountered today.

In the early days of Spaniel trials it was not unknown for a small Spaniel to win the Cocker stake and his larger litter brother to win the Springer stake but it seems that by 1915 the two breeds had become separately identifiable, as H. W. Carlton in his book 'Spaniels for Sport' definitely classes them separately. No other gundog breed has diverged so strongly and irrevocably into two types, the show and the working, as has the Cocker.

Field trial and show Springers are as different as chalk and cheese but can still be identified as Springers, whereas one could be forgiven for supposing that show and working Cockers belong to entirely different breeds. The show Cocker has a narrow skull, high domed head, ears set on low, very pendulous and pear-shaped. He has abundant feathering, a sloping topline and a deep muzzle with fleshy lips. I do not like him one little bit. A working Cocker is a very remarkable little machine. He has ears of very moderate length, often pointed. He has a flat and proportionately broad skull, on which the ears are set on high. He is often rather snipey in the muzzle, has usually rather large, prominent eyes and a longish reach of neck. He has a level topline and a very muscular body, particularly in the areas of neck, shoulders, loin and hindquarters. He is enormously strong for his size, stronger, kilo for kilo, than a Springer or Labrador, and one has only to see a Cocker racing back at full gallop with a wounded hare, to have some conception of his unique power. Such a performance is the equivalent of a 12 stone man, picking up a large roebuck in his mouth and running full speed with it.

I do not know if I like the working Cocker. Sometimes, I am sure, I loathe the little beasts but I certainly admire them. One of my female trial colleagues says some unprintable things about them but I can quote her as saying, 'They are full of character and it is not all good. You have only got to look at their faces, looking out from the back of a car and you have to laugh.'

When one reviews a breed one admires, it is all to easy to have to the fore of one's mind, the better specimens one has encountered and the more spectacular performances witnessed but, if one is to write a factual account on a breed, it becomes a matter of integrity to report the other side of the coin. When I witnessed my first Cocker Championship, shortly after the advent of myxomatosis, there were some very good Cockers running, in the hands of well-known trainers, dogs which had been trained and experienced before the disease struck and who, therefore had been accustomed to having plenty of rabbits shot over them in their formative days. From the 1956 Championship onwards the Cocker declined very rapidly. The great Cockers, of which there were several of the pre-myxomatosis period, were pensioned off. My first impressions of Cockers at this time were that they were amusing, often efficient but very naughty little dogs, ever ready to take a rise out of their handlers but fascinating to watch.

In the period that followed, the majority of Cockers presented were nowhere in the same class as their predecessors, although the odd first-class dog still appeared and swept the Cocker deck for a period or, alternatively, may have been in the wrong hands for full justice to be done to it.

24. The author's first Cocker, Lanegate Toby.
C. M. Cooke

The Springer Spaniel appears to have overcome the handicap of myxomatosis far better than did the Cocker and seemed to adjust to the new conditions with consumate ease.

I have written elsewhere that a good Springer should have the qualities of the Spanish fighting bull and the Zulu warrior. He should have the savage frontal attack of both and the blind courage and the unquestioning obedience to his handler that the Zulu had for his king and immediate superiors. In the case of the Cocker, it is not so much a question of the qualities he should have but those he does have. A good Cocker reminds me of a combination between a Corsican bandit and a seven-year-old dog fox.

Basically, the Cocker needs more game to get it going than the Springer requires. Their attitudes are so different. Order the Zulu to charge a line of machine guns and he would charge without question. The Corsican would decline and take the guns by night attack.

Similarly, the good Springer says, 'Here is a piece of country. Let us smash it apart and shift any game that happens to be there.' On the other hand the Cocker works the problem out in his villainous little mind and says, 'Here is a piece of country. Nine rabbits fed on the edge of this cover last night. Three pheasants used this track an hour ago, see, one has been scratching in those dead leaves; those gorse bushes look useful. I think we should investigate. This one smells good. This bush smells better and this one better still, as a rabbit is trying to hide in it. What abominable bad luck for him. What an enormous prank if we can frighten him out and how much more delightful if we can catch and kill him before he can escape.' This passage, of course, over-emphasises the potential evil of the Cocker to illustrate the point that the Cocker works best when his nose supplies him with direct evidence of the presence of game, rather than suggesting that he works in the hope of finding game, as does the Springer although some individual Cockers occasionally possess a more Springer-like attitude.

The Cocker has a terrier-like quality to his hunting. He can fit in anywhere where game can go and makes use of all the tracks and runways used by game just as a terrier would. A terrier may be quite happy to assault a fox in its den but does not believe in assaulting a solid wall of cover without an entry. So the Cocker tends to look askance at solid cover but show him the smallest entry and he will attack the bush right at its base. He detests cover which has been tampered with by man, and is at his worst in a plantation which foresters have trimmed out, leaving drifts of cut cover and the cut stubs of brambles and suckers to stick in his feet. In any case, his beloved rabbits are not normally present in such places.

He is at his best on a Scottish hillside, covered in whins and bracken and holding a healthy rabbit population. A good Cocker very seldom misses game. His nose tells him which kinds of game use the ground and when it is rabbity, he will regulate his pace accordingly. I have seen just a very few galloping flat catchers in the breed but the proportion is lower than in Springers. Most Cockers really mean to find game.

Towards the end of the last century and for the first decade or two of our own, several books were written on field sports and gundogs by authors who, for want of a better term, I will call 'gentleman writers'. The reasons are fairly obvious. They had, first and foremost, a keen interest in the subject, the leisure for writing and the education to convert their thoughts into the written word. More practical men, such as gamekeepers and professional trainers, lacked the education to record their experiences for posterity. Some of their works are very good and highly readable but occasionally such authors present us with misconceptions. One I have encountered states that a Cocker can creep into small places which a larger dog could not penetrate. This is completely untrue, as a big, high couraged Springer can get into anything he can get his nose into, using his weight advantage and drive from the haunches to effect penetration.

From a hunting angle, I do not believe a Cocker is ever at any advantage over a Springer of equal quality except on a warm day. This is a point I have never seen raised in print before but, according to my personal experience, confirmed by other Cocker colleagues also, a keen Cocker can work the best Springer into the ground in warm weather. I have seen a Cocker do a long, hot run in very heavy cover and finish the performance by galloping back with a hare, under conditions when a Springer would be puffing and blowing and fumbling his retrieve. This brings me to the point that one cannot assess a breed until one has shot over it extensively. One can watch Cocker trials and even judge them competently but it is impossible to appreciate their finer points until one has trained and handled them.

It has been said that Cockers are more difficult to train and more inclined to hunt for themselves than Springers. I cannot say that I have found this to be the case. Perhaps I have been extraordinarily lucky in the Cockers I have trained or perhaps Cockers have an affinity for me (rather than the other way about). When I consider the members of the breed which I have trained from scratch, I can say they have mostly been very straightforward and not particularly hard dogs, although they tend to be difficult to make sit and stay down. They are inveterate belly crawlers and the sight of one pulling himself forward by his elbows, hind legs stretched straight out behind him causes me such amusement, with a resulting breakdown in concentration, that the chances are he leaves the drop completely and flings himself all over me. It then requires a real effort of will to seize him by the scruff of the neck and carry him back to where he came from, dangling like a dead hare but with his tail going furiously. Psychologically, they are good dogs to train. They are so happy under training, so eager to come out of the kennel for a training session, they make me want to get on with the job. Every good Cocker I have trained has been a natural retriever, right up to hand with no fumbling. Oh, yes, I have had the odd dummy buried but I find that if I run up and catch him while he is digging and shake him up a bit and curse him, he associates the punishment with the last action performed, i.e. digging a hole for the dummy, and does not get the idea that the punishment is associated with the retrieve. I have, however, had experience of two Cockers who had originally been trained elsewhere and who had developed a head of steam. These were incorrigible with an unbelievable rock hardness. One was perfectly steady if I shot over him as a no-slip retriever. I shot pigeon all round him one day and one runner walked towards us and passed within a yard of his nose, but set him in motion, shoot a bird and he was unstoppable, even if only a yard from me.

The other was twice as bad. He is the only dog who has ever run in on me at a drive and the same day, walking some ground between drives, he even broke heel and chased a pair of grouse for a considerable distance. His working potential was fantastic. He was a big dog and could equal any Springer in brambles. His mouth was variable and his heart as black as the coal beneath his native Monmouthshire hills. His owner confirmed that he only understood the most brutal handling which neither of us were prepared to give. Truly, he was sired by one of Satan's paragons as his dam had a litter by another sire, which we both found turned out as easy and honest as Spaniels could possibly be.

A weak point in many of the breed is that they do not mark as well as Springers. This has nothing to do with defective vision but rather with attitude. A Cocker will often flush a bird then glance sideways instead of marking the fall. Some will mark a fall, then take their eyes off it. They make up in part, however, by remarkable gun sense and have an uncanny knack of getting out to a bird in wide casts, often picking it almost as quickly as if they had made a direct mark. There is, of course, an added danger of them bumping into live game when working in such a manner and it would improve the breed's reputation, from a spectator angle, if more Cockers would mark well. Very few Cockers, indeed, are anywhere near as good at taking directions on blind retrieves as are the majority of Springers and the sight of a Cocker making an unproductive effort on a simple blind retrieve in a trial, coupled with the lamentable fact that we have seen a good many weak hunters in the breed over the last fifteen years, have tended to bring the Cocker to a state of disrepute. The attitude of some Cocker owners has done little to help their breed for, instead of recognising and admitting their dogs' weaknesses, they have stuck their heads firmly in the sand and avowed that their dear little dogs were not only as good as Springers but even better. It is of the greatest assistance when training and shooting over Cockers to have one or two first-class Springers on the premises, so one can shoot over both breeds almost simultaneously. The one serves as a yardstick for the other.

I used to shoot over some very tough marshland on the Lleyn Peninsula at a time when I was experiencing a brace of first season Cocker sisters. I would hunt them, one at a time, alongside a tough Springer who would take the whole country in his stride. The little Cockers were certainly softer than the Springers and would go where the going appealed to them, coming in to heel when the going became particularly rough. From time to time, they would do fine individual pieces of work, finding hares and pheasants in competition with their bracemate Springer, or retrieving duck from over a wide drain. Later on I was to use, on this same ground, a Cocker who could take any ground we encountered and whom I rate as high as the best Springer in the country.

Of all the gundog breeds for which field trials are held, I would say the Cocker adapts to trial conditions less readily than any other breed. They hate waiting their turn on the lead and performance often suffers. Some-

25. A top American trainer, David Lorenz (left).

times at a trial for Cockers, a rot seems to set in, dog after dog going wrong and usually when someone is reporting the trial who is known to be notoriously anti-Cocker, yet many of us know that several of these dogs who have gone wrong have previously done sterling work and, at home, are excellent shooting dogs.

Most Cockers who are used for shooting and field trials are entirely working bred with two notable exceptions but more of them anon.

TRAINING THE COCKER

It is possible to write a whole book on gundog training and still not cover every point so I will not attempt to cover the whole syllabus of training a Cocker in depth. Before I trained Cockers myself, being a mere Springer and Retriever man, I listened in awe to the tales of the well nigh impossibility of a Springer man ever attempting to attain success with a Cocker, as there was some mystic art in the procedure which was only understood by Cocker men. One had to know the right incantation to be muttered at the correct phase of the moon and no Cocker trainer would divulge these secrets. I was told one must never hit them, or they would never work, neither must one allow a Cocker an inch of latitude, otherwise the Cocker would seize a mile and be lost for all time. Looking back, the strangest thing is that I attempted to believe these yarns!

When my first two bitches came to me, I was very careful to give them a very strict obedience course. Certainly I never hit them. I never had to. They were such lovely, gentle little creatures, so gentle in fact that it soon became perfectly obvious that under this system they would never hunt in their lives. What did I do? The same as I do with any Spaniel that shows no inclination to hunt. By this time it was early June and the bracken was just springing in my rabbit-pen, so in went the pups, every night and soon learned that it was great fun chasing rabbits. Night after night they hunted rabbits, chasing them at will until I could see that essential drive had developed, then I set about steadying them. They were, of course, fully trained to the stop whistle and steadied remarkably easily.

It must be difficult for anyone who has had trouble in the shooting field with their dog chasing ground game to believe how easy it can be to steady a Spaniel which has hitherto been allowed to hunt and chase at will. The secret of allowing a pup to free hunt is never to carry it to the point when the pup goes berserk. One must carefully assess the intensity with which the pup pursues a rabbit and discontinue free hunting directly the required level of keenness is reached. When Speckle of Ardoon came in for training at seven months of age, she was biting into cover like a veteran. The drive, style and desire were already built in, so I never free hunted her. Some courses after the Ulster hares had set her alight and she had learned to follow a rabbit through thick cover, so it became a matter of channeling what her breeder had already developed rather than developing it myself, as I did with my first two bitches.

It has been said by some authorities that it is wrong to allow a gundog to do wrong, then attempt to eradicate this wrong in future training. The point I wish to make is that it is not wrong for a puppy to chase and is the equivalent to the 'hacking' period of Peregrine eyasses, when they are allowed to fly free before training and develop a taste for pursuing birds. What is wrong is that a young dog should be allowed to chase in the shooting field and run in to shot game. I do not believe the pup should learn 'the power of the gun' before he is completely steady although many Americans would disagree. To encourage a direct flush and keen mark in their trial Springers, many Americans trainers shoot birds over their pups before steadying lessons have commenced. I do not myself agree with this system.

From the foregoing it will be seen that I believe in developing the natural instincts of a Spaniel before introducing discipline, particularly in the case of a Cocker, as weak hunting, in particular, has been a debit mark in the breed for so many years but again I stress the point that when a pup has become really keen, flattens his ears and pursues a rabbit completely oblivious of anything but the rabbit, it is time to call a halt to uninhibited hunting.

Before I steady a Spaniel to rabbits and game, I bring it to a fairly advanced point in its retrieving lessons for reasons which I will explain. At first, I allow all my pups to run in to a dummy to encourage keeness and marking. After a time, I will tuck a pup under my arm and throw the dummy into cover, the object being to prevent him being right on the tail of the dummy as it hits the ground, so when I release him, he has to run out and really use the ground and the wind to find his retrieve. The next lesson is steadiness to the fall. Completely open ground is best so he can see the dummy on the ground. One can sit the pup down, preferably in a narrow piece of ground where one can almost block the entry by standing in front of the pup, and throw the dummy over one's shoulder away from him. As he attempts to run in, he has to pass close to his trainer, who attempts to stop him by verbal command or whistle. If this is accomplished he should be kept on the drop for a few seconds then encouraged to fetch the dummy. Having been stopped a sensitive pup will often prove reluctant to go out and retrieve but verbal encouragement plus the sight of the dummy should cause him to go to it. If he succeeds in breaking past his trainer, the latter should pursue him as quickly as possible and snatch him up before he can touch the dummy. He must not be punished but carried back to his original spot and told to 'Hup' sharply. The dummy should be picked by hand and the process repeated until he will only retrieve at the handler's command.

A variation of this lesson is to put the pup on a lead or cord, preferably tied to a gate or post. There must not be too much slack on the lead as it is undesirable for the pup to be brought up with too hard a jerk. The dummy is thrown, the pup starts forward, the command 'Hup' is given (or stop whistle blown) just as the lead pulls him up short. The pup is kept sitting, the lead is carefully removed, the command 'Hup' given again and the trainer straightens his back. The pup is sent after several seconds as in the first method. I would recommend the latter method for a pup of more sensitive temperament but all the Cockers I have trained have been bold dogs by the time this lesson has been reached, unlikely to be put off by a rough character charging after them.

Do you imagine the pup is now steady to a thrown dummy? He is not. Set him off hunting, throw the dummy and he will run in to it as though he had never been stopped at all. The reason? He has been steadied from a sedentary position. His movement breaks his contact with the lesson and in he goes. He must be re-steadied whilst in motion. The wise trainer weighs his opportunity, throwing the dummy when he is in an advantageous position in relation to the pup, to intercept and stop it when it tries to run in. Here I utilise the steepness of my ground. I throw the dummy uphill when the pup is downhill from me, as he will be far easier to stop when he is travelling uphill than on the level.

I give him a good deal of this type of work, sometimes dropping the dummy close in front of him until he will actually take the falling dummy

26. A study in concentration. Jack Chudley with F.T. Ch. Exton Dainty.
Peter Moxon

27. Mrs P. Leigh's F.T. Ch. Goldspan Bray, winner of the Cocker Championship at Cornbury in 1969. A good-looking working Cocker.
Peter Moxon

as a signal to drop with no word from me.

Next I introduce the gun. This will either be a ·410 or a converted Martini carbine which takes a ·22 blank and which is so designed that it will take the place of a dummy launcher. I seat the pup and walk about 100 yards away, with the wind blowing from him to me. I fire a shot and call him up to me by whistle, observing his reaction from the second the shot is fired to the time he comes up to me. He should remain perfectly still to the shot, his ears remaining pricked forward and stationary and, on recall, come tearing back with tail going. I repeat this at decreasing distances, finally firing a shot when he is 25 yards away, throwing the dummy and letting him retrieve it and gradually work in the idea that something pleasant, i.e. a retrieve, is related to a bang.

I have never had a gun-shy Cocker, that is, one that runs and hides at a shot and refuses to work thereafter, nor have I had a truly gun nervous one—one which flinches at the shot, showing fear but will remain with the handler and make a retrieve afterwards, probably with ears and tail down and maybe not coming up to hand so boldly as is his usual habit. But I *have* had one of those deceptive individuals in the breed, the dog which does not care much for a shot. He does not flinch at the shot, he comes boldly in afterwards and a trainer has to be extremely sensitive indeed to spot that anything is amiss at all. He may just duck his head slightly in anticipation of the shot as the dummy leaves the trainer's hands. Upon retrieving, he might attempt to circle a little and turn his head away. Failure to recognise the signs and continued shots in his presence could cause him to become truly gun nervous. This condition is, I am quite certain, responsible for the dog whose trainer tells one, 'He was perfectly alright with a gun then suddenly developed gun nervousness.' The trouble was always there but the trainer did not recognise it.

I once asked a colleague how a certain Cocker was progressing and was told, 'She's a silly little thing. She's not gun nervous but if you fired twenty shots over her in a grouse butt, she would be.' This was said by a man of above average dog sense. In my own case, I carried on with seating the dog at a distance and firing a shot but each time he came up to me I fed him a small piece of tripe and progressed until a shot right over his head was a signal for him to look for something in my hand.

Once the dog is retrieving well at my command only and is perfectly sound with a gun I establish his confidence in retrieving a live bird. He will already have handled dead ones, as and when available, from being quite a small pup. My Cockers have all been bold natural retrievers, capable of handling a dead cock pheasant at six months of age and never flinching at a bird as do some Springers because it is an unaccustomed object. I have never hit a snag when allowing a Cocker to handle its first wounded bird and once I can see this is well established I steady the pup to flush. I never do this until he has handled live game, as an intensive course of steadiness to rabbits can cause him to eschew live game later and although this can eventually be overcome, a good deal of work can be saved if the trainer establishes his confidence in retrieving live stuff before the pup is steadied to flush.

Cockers can be quite hard to steady but are normally of resilient temperament, so one can wear them down with no loss of drive. I simply blow my stop whistle the moment the pup finds and attempts to chase a rabbit. Occasionally a pup will stop at the first time of asking if he has had a good grounding with the stop whistle but usually he will ignore it and pursue the rabbit. The trainer must chase him, head him off and catch him, giving him a shake and a sharp blast of the whistle in his ear. Punishment beyond a shaking should be avoided at this stage as the object of this lesson should be to show him he must no longer chase, rather than punish him for chasing.

Once a dog really knows right from wrong, has been shot over and any deviations from good behaviour been corrected by the vigilance rather than the severity of the handler, should he then blatantly misbehave, he can be dealt with more harshly. Experience in the field toughens a dog mentally, so he should accept punishment at this stage for really serious wrongdoing, such as running in to shot game or chasing, punishment which I would not think of advocating during the training course. However, if one is very careful during the transitional period between training and early shooting days, a good handler can usually 'bed a dog in', in such a manner that it remains steady for the rest of its life.

The basic principle is to watch for and anticipate the least sign of unsteadiness to game. If the dog is tearing through a gorse or bramble bush, with the flush of a rabbit or bird imminent, the stop whistle should be blown directly game and dog break cover. He may stop on his own but the whistle will remind him that his handler is still in control. With more experience, he can be trusted to stop to flush on his own. He may flush a pheasant in a fairly open situation, the bird being shot in full view. As soon as the bird is shot, give the command 'Hup', and reload the gun. Keep him waiting 30 seconds then send him for the retrieve. If in this waiting period he attempts to retrieve on his own, stop him. Watch him carefully after the shot. The time to stop him is not when he has gone a yard but when you see his muscles bunch to go.

Try to develop a reaction swift enough to work as soon as head and shoulders sway forward to go. Use a good loud 'Hup' or the stop whistle and you will hold him. Remember, a dog's first run is seldom a real full-blooded, unstoppable charge. Having held him, put your gun down, cross swiftly over to him and give him a good cursing if he is a sensitive dog and if you know he is tough, grasp him by the muzzle in one hand and an ear in the other and give him a single shake. When Spaniels fight they very rarely fight to kill, although bitches in season can prove the exception to the rule. Most Spaniels fight a good deal to establish their supremacy but not to kill the opposition. The target, therefore, is seldom the throat but the lip, foreleg and ear, so the handling of a dog's ear establishes the handler's supremacy over it and puts him in the position of leader of the pack. Having shown the young dog his error, the handler should pick the bird himself and try to shoot another one. Failing which he should fire a shot and throw the bird out, make the dog wait and then send him for it and so illustrate that he still expects the dog to retrieve on command. These tactics have kept me well to the fore in the competitive sphere of Spaniel work for fifteen years. Go thou and do likewise. No mystic art is required.

FEEDING AND AVAILABILITY

Do not be misled by the small size of the Cocker into believing that he will need less food than a larger dog. Certainly, the summer maintenance diet of a mature Cocker not actively engaged in any work can be less than that of a Retriever or Springer but when he is working hard in the shooting season and burning up a lot of energy, the difference in the amount of food that the Cocker and a Labrador will require will be minimal. I feed bullock's tripe, not too well cleaned and a well-known brand of complete dog food, Solo, which is manufactured by some keen Cocker supporters. I feed far more tripe than Solo to my adult dogs, but increase the quantities of the latter for bitches in whelp as a morning feed. For puppy rearing, I place a hopper of Solo High Protein Meal in the kennel when the pups are 3 weeks old and allow bitch and pups to help themselves. Solo is thirsty stuff, so water must always be available. By the time the pups are 5 weeks old, I give a small feed of honeycomb tripe, cut into one inch squares, every night. I am a great believer in raw tripe. It is high in protein, though said to be a little short in minerals and keeps dogs well muscled but lean and healthy.

To acquire a pure working Cocker is not always an easy task at best and at worst can prove an extremely hard one. My advice to any prospective purchaser is not to rush in and buy from any person who is simply known to own field trial Cockers. He could well be buying from a kennel which has consistently turned out the weak hunting type which has harmed the breed in recent years. The potential Cocker owner should study field trial results for a season in the sporting journals such as *Shooting Times and Country Magazine*, *The Field*, or *Gamekeeper and Countryside*. He should take note of those dogs and owners who have been doing consistently well and make his needs known to one or more of them.

It is virtually impossible to go to a Cocker owner and simply buy a pup from him. It is advisable to approach him many months in advance and order one, possibly before the specific bitch has even been mated. If one breeder cannot supply a pup, as he may wish to run all his pups on for his own use, he may be able to advise an enquirer as who may have some for sale, or who has recently used his stud dog and may have a pup in a few months time.

Working Cockers are a minority breed and seldom have large litters but the more interest that is shown, the more pups breeders may be induced to breed, over and above their own needs. The working Cocker is in a cliff-hanging position. There is a nucleus of very fine stock in the country but a shortage of patrons willing to put money into the breed by having Cockers professionally trained and run in trials. There are signs that interest is increasing slightly but we need more Cocker owners, so when a good dog appears, he can mate many more bitches than is the case at present. If the breed can become stronger, both in quality and numerically, Cocker pups will become easier to obtain. There is one advantage on the Cocker's side. The English Springer is becoming weaker. There is now a working test element which is bringing the wrong class of dog to the fore, which can only result in harm to the breed.

When looking at the pedigree of a Cocker, some good prefixes to look out for are as follows:—Elan, Stockbury, Deewood, Carswell, Cromlix, Staxigoe, Lanegate, Templebar, Merlin, Glennewton, Exton, Henhamlodge, Monnow, Headland, Greatford, Chrishall, Breckonhill, Jordieland, Debdenhall, Brookville, Shaffylde, Ardoon and, I hope, Gwibernant.

Winners of the Cocker Championship since 1930, have been:

1930. Mr H. S. Lloyd's bitch F.T. Ch. Tiptoe of Ware
1931. Mrs V. Berdoe Wilkinson's dog F.T. Ch. Michael of Silverlands
1932. Lorna, Countess Howe's bitch, F.T. Ch. Banchory Rachel
1933. Mrs V. Berdoe Wilkinson's F.T. Ch. Sue of Silverlands
1934. Mr J. H. Gibson's dog F.T. Ch. Heath-Mynd Nap
1935. Mrs V. Berdoe Wilkinson's F.T. Ch. Sue of Silverlands
1936. Mr G. Clark's dog F.T. Ch. Judy of Blair
1937. Miss J. Clark's dog F.T. Ch. Punch of Blair

1938. Miss J. Clark's dog F.T. Ch. Punch of Blair
1939-1948. No Trials held.
1949. Major G. Roslin-William's dog F.T. Ch. Lash of Mansergh
1950. Mr J. Kent's dog F.T. Ch. Newton of Chrishall
1951. Mr R. B. Weston-Webb's dog F.T. Ch. Meadowcourt Breckonhill Beau
1952. Mr J. Scott's dog F.T. Ch. Shawfield Glenfire
1953. Lt. R. H. Ogden's dog F.T. Ch. Merlin Nicky
1954. Mr G. Fiske's dog F.T. Ch. Debdenhall Timothy
1955. Lt. Cdr. Collard's dog F.T. Ch. Buoy of Elan
1956. Major H. Peacock's bitch F.T. Ch. Merlin Minimus
1957. Mr J. H. Windle's bitch F.T. Ch. Jordieland Bunty
1958. Lt. Cdr. Collard's dog F.T. Ch. Buoy of Elan
1959. Mr H. C. Martineau's dog F.T. Ch. Brookville Sandy
1960. The Countess of Portarlington's F.T. Ch. Greatford Crick
1961. Lt. Cdr. Collard's dog F.T. Ch. Simon of Elan
1962. Lt. Cdr. Collard's dog F.T. Ch. Simon of Elan
1963. Mr J. W. Smith's bitch F.T. Ch. Henhamlodge Dynah
1964. Lt. Cdr. Collard's bitch F.T. Ch. Bliss of Elan
1965. Miss V. M. Pinney's dog F.T. Ch. Headland Irish Moss
1966. Lt. Cdr. Collard's dog F.T. Ch. Jet of Elan
1967. Lady Auckland's dog F.T. Ch. Wilfred of Cromlix
1968. Mrs P. Leigh's dog F.T. Ch. Goldspan Bray
1970. Mrs T. A. Watt's bitch F.T. Ch. Exton Jenny
1971. Lady Auckland's dog F.T. Ch. Jimmy of Elan
1972. Mr W. C. W. Sloan's bitch F.T. Ch. Speckle of Ardoon
1973. Mr K. A. Erlandson's bitch F.T. Ch. Speckle of Ardoon
1974. Mr K. A. Erlandson's bitch F.T. Ch. Speckle of Ardoon

F.T. Ch. Jordieland Bunty

Bunty won the first Cocker Championship I ever witnessed, at Grimsthorpe in Lincolnshire. She was a very tiny but compact bitch, blue roan with a good round skull and solid muzzle. To offset her diminutive stature, she had a habit of standing on her hind legs at a shot, the better to mark.

During this Championship, she gained distinction by making a good retrieve of a runner upon which a previous dog had failed. She made one clever retrieve of a dead cock which snagged on a bramble. Bunty whipped round in her own length and disengaged the bird in an instant, continuing her journey completely invisible behind the bird, which appeared to come in to the line with a propulsive force of its own.

Bunty's owner was a gamekeeper in Galloway, so she was brought up as a practical working dog. Her progeny and descendents have hardly figured at all in trials, being mainly used as shooting dogs. She died fairly recently, aged 17.

F.T. Ch. Buoy of Elan

Buoy was a very well made black and white dog who, besides winning the Championship in 1955 and 1958, was also Reserve in this event in 1957. Upon the later reflection of his handler, the late John Forbes, who had several successful dogs following Buoy, this dog was still rated as the best he ever had. He was a particularly kind and easy dog to handle, never hotting up on the lead during competition. He was also good at handling out to his blind retrieves, unlike many. During the 1957 Championship, an American guest gun claimed a right and left at woodcock, both of which Buoy was asked to retrieve. He collected the first in fine style, handled out well to the second, then appeared reluctant to pick it and upon receiving encouragement from John, picked it up with a shake of his head and delivered—a little owl!

F.T. Ch. Galbry Daniel

Daniel was a very big dog, black, white and tan, who perhaps did not have the true Cocker action but who could really 'go'. He is reputed to have won four stakes during the course of one week and could probably have won a Championship but his handler judged the event that particular year, so Daniel was handled by his owner, who had never previously handled in a trial and lacked the experience to keep Daniel working to a consistent pattern. Nevertheless, Daniel, working entirely on his own, ran a very good trial, collecting a very good runner and finished Reserve. A most commendable performance.

F.T. Ch. Simon of Elan

Twice winner of the Championship, Simon was a short coupled blue roan. He had a lot of force but was more of a handful than Buoy. Simon really taught me what a true Cocker action really is. In one Championship he ran off against a superbly trained little dog but which, in action, resembled an inferior little Springer; this dog had a very good retrieve in the course of the run-off but was run off his feet by Simon. His handler appeared rather disappointed by the result but, to an informed onlooker, there could be no question of the superiority of Simon's hunting.

F.T. Ch. Wilfred of Cromlix

Sired by the famous F.T. Ch. Merlin Micky out of his owner's Nix of Cromlix, Wilfred was truly one of the great ones. He was a big Cocker but not coarse, with a rather long body and, when hunting flat out, the long raking gait of a Springer. Wilfred had the advantage of an ample game supply throughout his career, which no doubt helped to develop his quality. He really could find his unshot game, going at all times with great drive and could use his nose and brains to get his shot game into the bag and was particularly good on a runner. He left some successful triallers behind him, F.T. Ch. Templebar Blackie, F.T. Ch. Glennewton Keeker and F.T. Ch. Tayburn Ron, all of which have won awards in the Cocker Championship.

F.T. Ch. Pinehawk Rebb

Rebb was a particularly brilliant dog and kind as well. He only ran in trials one season, as a puppy, winning his first Novice Stake at Cromlix and subsequently won six Open Stakes. He had to forgo a run in the Championship the year Wilfred won it as his owner, Jimmy Wylie, again judged the event.

Rebb was liver and white and was sired by F.T. Ch. Swift of Elan out of Bright of Elan. As far as I know, Rebb went abroad, leaving very little stock behind him. He did, however, sire my first two bitches with which I competed in trials. The better one, F.T. Ch. Monnow Mayfly, ran very consistently and won 3rd and Reserve in Cocker Championships. She is now in Texas, a fact which causes me some guilt, when I consider the quality of the only litter she produced.

F.T. Ch. Templebar Blackie

Blackie throws back in appearance to his grandsire F.T. Ch. Merlin Micky. After a shaky start in trials, when he was prone to disappear in the wake of the big Galloway hares, Blackie settled down under the guidance of his handler, Rosemary Hall and became one of the most consistent Cockers ever. His strongest point is his methodical hunting pattern. He covers every inch of the ground and never attempts to pull out of shot, working as though attached to an invisible piece of elastic. He is particularly good at poking rabbits out and is very good on a legged rabbit or runner. He has won seven Open Stakes and has been 3rd in the Championship.

He seems to be passing on some of the prepotency of the Merlin Micky line and has sired two field trial champions in one litter, Templebar Tammy and Ardnamurchan Mac, besides some other promising young stock.

F.T. Ch. Speckle of Ardoon

The telephone rang and a well-known Irish voice said: 'You remember that Cocker wee Bob had?'

'You mean the one that won't work.'

'Sure. You remember the old Consul I had?' (I didn't. I lost track of his various old bangers).

'No.'

'Oh well, it doesn't matter. I swapped it for the Cocker.'

'Oh, really?'

'Sure, and wee Bob threw a Springer into the deal as well.'

(Interest now awakens) 'What's it like?'

'Too damned steady.'

'Will it work?'
'Sure.'
'Send it over, I need one for a man in South Wales.'
'The Cocker's in season. Who do I mate it to?'
'The only working Cocker in Ireland is her own brother. You'll have to take her to a Springer. Use Goldeneye Jock, he's a small dog.'
'I'll do that, I'll do that.'
Several months passed. The phone rang again. 'I took that bitch to a show Cocker down the road, she had two pups. I've kept a bitch. Greatest thing you ever saw.'
'I'll bet.'
'Best thing ever. Better than any of my Springers. I'll send it over. You can train it.'
'Alright, if it's that good, send it over but I don't believe you.'
He did send it over. It was the greatest thing I had ever seen and the way it was bred, it should not have been—by a show Cocker out of what was virtually a dud, of trial breeding. Everything was all wrong. Surely her breeder must be very well in with the small people.
He rang a little later. 'How's that article.'
'Oh, I'll make it a field trial champion.'
'Ah sure. Do that, do that.'
Well I did just that. Speckle ran in a Novice Stake before she ever had a single head of game shot for her and received a certificate.
I shot three head of game for her, a hare, a strong running cock pheasant and a various, then won a Third in an any variety stake with her. She ran a good second to Ardnamurchan Mac when he won a Novice Stake, then

28. The author's F.T. Ch. Speckle of Ardoon. The most consistent winning Cocker for 20 years—perhaps of all times.
Peter Moxon

she entered open competition and won four Stakes in her first season. By the time she won her second Stake and gained her title of field trial champion, she had retrieved precisely 18 head of game, which proves that a dog of exceptional quality does not require a colossal head of game to be shot over it before it hits winning form. In three seasons she has won nine Open Stakes and three Cocker Championships, a record, but I am sure that she will be remembered more by the ordinary shooting man for winning the Spaniel Tests at the 1972 Game Fair at Raby Castle. I do not think any Cocker has ever qualified for the final of this event before, let alone won it, as even good Cockers do not generally lend themselves to this type of work.

Instead of two heats of 4 or 6 dogs each, the best two to be taken from each, which gives one a better chance of qualifying for the final, should one find one's dog in the weaker heat, we had one heat of 12 dogs, the best four to qualify for the final. Speckle won the entire heat by one and a half points from F.T. Ch. Harwes Silas, a particularly big Springer who won the previous year. For the final on the following day, the sun came out with a vengeance and I knew we had the three finalist Springers beaten. Speckle can run any Springer into the ground on a hot day and this she did, increasing her lead by three points over the next contender.

She has an almost feral quality to her hunting and we who know her well usually refer to her as 'The Polecat'. Considering her breeding, she is a freak of nature but has a brother in Ireland who, I am told, is even better and is simply used as a shooting dog. I discussed the question of their breeding with an ex-Master of Foxhounds, suggesting that the good

29. Judges closely watching delivery at a Cocker stake.

working blood on her dam's side had become so close that the bitch became inoperative and the show blood from her sire diluted the dam's genes by 50 per cent and gave them room to swim around again. After many years experience of foxhound breeding, in which similar things had happened, he was inclined to agree. I was also told by a lady of considerable experience that the situation was probably aided by the fact that the sire, Tireragh Silver Starlight, was of predominantly sixshot breeding and, for some unknown reason, sixshot blood is still capable of passing on working qualities.

What is to happen to these fascinating little dogs whose ancestors roamed the hills and marshes with Hywell Dda and are surely part of our sporting heritage? That there are good Cockers who can do a useful day's work is a fact beyond all doubt but no breed can survive without support. We need more Cocker owners who are willing to use them as shooting dogs and to finance them in competition. The breed at the moment is being kept alive by a handful of dedicated supporters but their real future lies in the hands of you, the reader.

7 Irish Water Spaniel

The Irish Water Spaniel is the all-round shooting man's Complete Companion. This claim is made for many breeds but it is not always easy to justify. In the case of the Irish Water Spaniel actions speak louder than words and no one who has been privileged to work with these dogs in varied conditions can have any doubt as to their all-round suitability. The wildfowler who may spend hours flat on his face in the mud stalking geese, the keen snipe shot who walks miles of bog, the man who finds his pleasure in working heavy cover for woodcock, the Saturday pheasant shooter and the dawn duck flighter will all appreciate that devoted, skilled, intelligent accomplice, companion and friend, the Irish Water Spaniel.

This versatile animal of unusual appearance and strange shambling gait can be trained to perform a remarkable variety of tasks. Enormously hardy, his thick, curly brown coat enables him to withstand extreme cold; a strong swimmer, equally at home in the water as on the land; possessed of great intelligence, an excellent nose, good eyesight and, above all, a tremendous sense of humour, he will hunt silently, seldom leaving anything undisturbed and is undeterred by the densest cover though he may emerge with a high percentage of it adhering to his coat.

He is large enough to retrieve duck and geese from land and water and most owners relate tales of fringe activities which add enormously to the pleasure of owning an Irish Water Spaniel. A puppy I bred and sent to M. Robert Borde of Entraigues in the South of France not only swam the Rhone in spate but was also trained to hunt for truffles. His mother, Canon Kate, always helped me with the daily collection of eggs, picking one from the nest and carrying it indoors held delicately between her front teeth and in eleven years she dropped only one. Major de Mey, writing in the Irish Water Spaniel Year Book of 1972, tells of his dog which points and retrieves trout for him.

To live in partnership, for that is what it is, with an Irish Water Spaniel is, in my view, an essential factor of the quality of life. His loyalty to the family he owns is complete, making friends with strangers only slowly and never before their credentials have been satisfactorily established. Despite his curious appearance, he has great dignity and an innate sense of propriety; it is a brave being, human or canine, who risks any breach of good manners at home or out shooting, for it will certainly not pass without comment. One old bitch not only stopped the children squabbling but also reproved her master if he missed a shot. It is his ability to be involved in every part of one's life that makes him the complete companion.

HISTORY

The origins of the Irish Water Spaniel are difficult to trace with certainty, although most writers describe it as being one of the oldest breeds, thought to have originated in Persia, making its way via Spain to Ireland. Shakespeare refers to water-dogs in 'The Two Gentlemen of Verona', Gervase Markham, writing in 1595, mentions a liver-hued water-dog, and there is certainly a brown dog, recognisably an Irish Water Spaniel complete with top-knot and rat-tail, in the middle of a Gobelin tapestry of the time of Francis I, which now hangs in an art gallery in New York.

However, it is generally accepted that the Irish Water Spaniel as we know it to-day, distinct from any other water spaniel and unique in coat, colour, top-knot and tail, was firmly established in Ireland by Justin McCarthy during the 19th century and influenced particularly by his great stud dog Boatswain, who was born in 1834, lived to the great age of eighteen and whose blood could probably be traced in many pedigrees to-day. Bitches were brought to Boatswain from far and wide and Mr McCarthy bred Irish Water Spaniels for a great many years, sending puppies all over the world.

The description of the breed, which he gave in a letter to *The Field* dated February 19, 1859 in answer to a large correspondence, still holds good to-day. 'Dogs from 21 inches–22 inches high, head rather capacious, forehead prominent, face from eyes down perfectly smooth, ears from 22 inches–26 inches from point to point. The head should be crowned with a well-defined top-knot, not straggling across like the common rough water dog but coming down in a peak on the forehead. The body should be covered in small crisp curls which often become clogged in the moulting season. The tail should be round without feather underneath, rather short and as stiff as a ramrod: the colour of a puce liver without white.'

Although never a 'popular' dog, which is probably its greatest strength as it has never been indiscriminately bred, it has always had its devotees among sporting people and breeders, who recognised its qualities and remained faithful to it all their lives.

The Irish Water Spaniel was immortalised by that brilliantly humourous pair, Edith Somerville and Martin Ross, in 'Some Experiences of an Irish R.M.' first published in 1899, which every prospective owner of an Irish Water Spaniel should read, not only to become acquainted with that remarkable bitch Maria, given as a wedding present to the R.M.'s wife Phillipa 'to be a lamb at home and a lion in the chase' and who 'could be found assiduously licking the barrels of my gun as was her custom when

30. Zanfi Redstart, owned by Mrs Beavis.

the ecstasy of going out shooting was hers.' but also for the description of the lighter and softer aspects of life in Ireland at the end of the last century.

The Irish breeders founded The Irish Water Spaniel Club in 1890. In England in the same year Mr Trench O'Rorke of Bodmin in Cornwall became interested in the breed and started his famous Briefny line which he showed successfully at Cruft's and in Dublin; he ran in field trials and exported to many parts of the world right up to the last war. His Champions, Briefny Count and The Rose of Sharon, were successful on the show bench and in the spaniel field trials of 1924. In 1926, when the Irish Water Spaniel Association was inaugurated at a meeting in Birmingham, Mr

O'Rorke became its first Hon. Secretary and remained in office until the outbreak of the last war.

During the latter half of the nineteenth century Irish Water Spaniels were exported to the United States and by 1880 had become the third most popular sporting dogs due to their ability to stand extreme cold and the fact that they could be guaranteed to 'put meat on the table', a very necessary requirement in those days.

By 1900 Mr Percy Swan of Chico, California, was breeding a large number of dogs whose working ability was paramount and consistency of type remarkable. During the 1920s and 30s there were very large kennels in the States having up to one hundred dogs at a time. Mr and Mrs Goodnow were the mainstay of the breed with their Shillalah strain at their Princess Colleen Kennels. In 1937 the Irish Water Spaniel Club of America was formed and today has well over a hundred members.

The breed flourished in Ireland and the United Kingdom between the wars. In Ireland Miss F. Barrington of Lisnagry near Limerick with her Annagh strain and Mrs G. M. Barrington of Nobber, Co. Meath, present President of the Irish Water Spaniel Association, with her Brittas line were particularly prominent. In England several breeders produced excellent dogs; among others Mr and Mrs Groves started their Croome Kennels in the late 1920s and bred successfully until well after the war, keeping at times up to 30 dogs. Their Champions, Croome Teddy Boy and Croome Molly, had a distinct influence on the breed.

Since the war the Irish Water Spaniel has quietly maintained its popularity among its devoted followers. The I.W.S. Association keeps interest in the breed very much alive. Thanks to a succession of energetic Hon. Secretaries, first Dr. James, then Lt. Col. G. Thompson, followed by Mrs Harvey-Kelly and currently Mr Nigel Waters, members are kept well informed of shows, trials, availability of puppies and all matters concerning Irish Water Spaniels. A news-letter is circulated annually and since 1970 an excellent Year Book has been produced full of information for the addicts.

The Association has run its own field trials and provides several trophies. In 1968 the Shannon Trophy was given by Miss F. Barrington to be awarded annually to the best Irish Water Spaniel of the year, on a points system based on successes gained through the year at shows and field events, double points being given for the field events. This trophy has been won by some first class dogs; the first holder was Mesdames Gordon and Mackensie's Gormac Reed Warbler, followed by Mr and Mrs Waters' Sh. Ch. Leighbridge Wood Sage. In 1970 and 1971 the trophy was won by Mrs Harvey-Kelly's Tara of Tarbay and in 1972 by Mrs A. McDermott's Kitebrook Arran.

Although the Irish Water Spaniel is not seen in great numbers in field trials—it has to run in spaniel stakes to which it is not altogether suited—there have been some very notable successes. Miss Long's great dog Ch.

Seamus of Seedhill (bred by Mrs V. C. Yates, the vice-president of the Association) ran very well some years ago and now her Tallulah of Tarbay is following in his tracks. In 1971 Mrs Harvey-Kelly's Ch. Tara of Tarbay won the trophy for best performance at the Show Spaniel Field Day and in 1972 Mrs McDermott's Ch. Water Rail Dawn won the same award.

As an example of the great concern felt by members of the Association to maintain the standard of the breed, when Miss F. Barrington died in 1968, Mrs Harvey-Kelly, on behalf of the Association, made arrangements to fly eight of the Annagh dogs to England to ensure that this wonderful strain should be preserved. Mrs Sutton of Camberley and Dr. Rickards of Windsor kindly kennelled the dogs; one bitch promptly produced eight puppies and by June 1969 all but one Annagh dogs had found permanent homes.

On July 21, 1973 the Association held its own Open Show and County Fair at Kitebrook House, Moreton-in-Marsh, which was an unique event for a breed association of only one hundred members. The show judge was Miss Ank ter Kuile who has been largely responsible for the growing popularity of the breed in Holland.

Today the Irish Water Spaniel is in good heart. There is a growing interest in the breed, excellent strains of proven working dogs exist and the demand for puppies exceeds the supply.

DESCRIPTION

The Irish Water Spaniel is the largest member of the spaniel clan. The dogs stand between 21 inches and 23 inches and the bitches 20 inches and

31. Four Irish Water Spaniels from a photograph in *The New Book of the Dog*, Vol. 1 by Robert Leighton.

21 inches though they vary considerably in size, and weigh around 55 lbs. The gait is peculiar to the breed, sometimes described as shambling and somewhat similar to a bear's movement, yet it has a fluid attractive quality. The coat is composed of dense, tight, crisp curls of liver-coloured hair which has a natural oiliness. The forelegs are covered in curly or ringletted feather to the feet though that in front is short, giving a wavy appearance. The back legs are also covered in ringlets and below the hocks they should be smooth-haired in front but heavily feathered behind to the feet. The face is smooth-haired from in front of the ears but the skull, which is high and domed, is covered in a distinctive top-knot of curly hair, rather like a large chrysanthemum, which comes to a well-defined peek between the eyes. The muzzle is long and somewhat square with a pronounced stop, deep mouth opening and lips firm in texture, and there should be no snipiness. The eyes are not large but should be bright and lively, hazel or brown, the darker the better. The ears are very long and heavily leathered—covered in long curls—and they should be set low, hang close to the cheeks and reach to the end of the nose when extended forewards. The neck is long, strong and arching, carrying the head proudly well above the shoulders. The back and sides of the neck are covered in curls but the throat is smooth-haired with a clean-shaven V-shaped patch reaching from beneath the chin to the breastbone. The shoulders are powerful and sloping; the chest deep with well-sprung ribs and barrel-shaped. The body short-backed, broad and level, strongly coupled to the hindquarters; the forelegs well-boned and straight, the hind legs powerful with well-bent stifles and low-set hocks. The feet are large, well covered in hair even between the

32. Drenched to the tip of its rat tail but still alert and eager.

toes. The rat-tail, which together with the top-knot is such an unique feature of the Irish Water Spaniel, is short and straight, thick at the root and tapering to a fine point. It is low set and carried well below the back and should not quite reach the hock joint, the first three or four inches being covered in curls which stop abruptly while the rest of the tail is covered with very short straight hair. The colour of the coat should be a dark liver brown with no white at all.

CARE

Because Irish Water Spaniels are such intelligent and affectionate animals I prefer to bring them up with me in the house rather than outside in a kennel, so that the establishment of mutual confidence and trust is quickly achieved. No doubt some people keep their dogs successfully in kennels but they are not suited to long hours alone being naturally sociable, full of fun, interested in everything about them and enjoying the company of their family. They are most fastidious and therefore easily house-trained. Although their coats have a distinctive aroma it only becomes apparent if the coat is neglected and allowed to mat badly. We have always kept our dogs indoors with no discomfort on that score. If kennelled, a large bed should be provided, draught free, raised from the ground and filled with clean wheat straw which must be frequently changed. Indoors, I prefer the metal-framed bed with canvas lining for which I manufacture a mattress of several layers of towelling quilted securely together. This is easily washed. Inevitably, with the heavy coat, mud is a problem but I find my towelling quilts are the answer.

Feeding

The Irish Water Spaniel is a large dog, hardworking and energetic, and must be fed accordingly. Making do on scraps will not suffice. The eight-week-old puppy requires several small meals a day and the daily diet includes about 6 to 8 oz. of minced meat, well soaked puppy meal, an egg, milk or milk powder to provide enough calcium to form the fast developing bones and teeth, cod liver oil and yeast tablets. I always keep a stockpot 'on the go' to provide bone stock with which to soak the biscuit. I have no problems with uneaten food. My usual method is to give a soaked biscuit meal on waking, milk at 11 a.m., the meat at 1 p.m., milk and egg at 4 p.m. and a final biscuit feed at bedtime. This routine is reduced to three meals a day by about four months and to two meals at about six months, increasing the quantity of meat and biscuit as the puppy grows. When fully grown and mature I give one meal a day at 5 p.m., or at the end of the working day, which should include about 1 lb. of meat. A good supply of large marrow bones helps to keep the teeth in good order and provide an occupation when alone. I never feed small cooked bones or carcases.

The Irish Water Spaniel is easy to feed but it must be stressed that cheese and chocolate have to be kept under lock and key as neither commodity can be resisted. My old bitch, who could be safely left in charge of the Sunday joint, once consumed a nearly complete Brie when we were visiting my in-laws for Christmas; it is a measure of her tremendous personality that we only blamed ourselves for providing the opportunity.

Coat

There is no baulking the fact that the coat needs care. It must be combed regularly and kept reasonably free from tangles. If neglected, heavy mats of dead hair form which can only be removed by cutting away. The coat is naturally oily giving the water-proofing which enables the dog to withstand extremely severe conditions, and it is not advisable to shampoo it frequently. The dog's natural love of the water ensures that it keeps clean but, of course, when showing it is as well to shampoo at least a week beforehand so that the hair has time to settle down and recover its condition. The day before the show a rinse in water to which a little oil has been added helps to restore condition, and at shows I have seen startled owners of more delicate breeds staring in amazement at an Irish Water Spaniel having a bucket of water tipped over it to crisp up the curls—the coat certainly looks best when damp.

The Irish Water Spaniel does not feel the cold but, nevertheless, when the day's shooting is over the dog should be given a good rub down before being put back in the car. We usually use a brake and fill the back with straw which is a great comfort to the tired dog on a long journey home. They love snow but it sticks to the legs and tummy and must be removed with warm water before coming indoors. Curiously, the dense coat does not seem to be an inconvenience in summer, though oat grass can be troublesome as it burrows into the long hair and can even penetrate and enter the skin, sometimes emerging in another place altogether. We have avoided this unpleasant business by clipping away the hair from under the arms, underneath the ears and inside the ear itself and also cutting away the hair from between the toes where it is quite abundant and forms into small mats. If the dog is to be shown during the summer this must be done carefully so as not upset the look of the feet but it is a wise precaution as very unpleasant inflammation and sores can be caused by oatgrass.

Besides cutting the hair on the inside of the ear during summer, ears should be regularly inspected and kept clean, as constant swimming in ponds and rivers enables all sorts of mites to enter. It is wise to ask the vet. for a preparation for routine use.

The feet are strong and tough and stand up to days in the heather but they should be carefully inspected after shooting as they sometimes get pieces of bramble or thorn stuck between the toes which causes soreness. The dewclaws and toenails should be kept short.

WORKING ABILITY

I have used Irish Water Spaniels as working companions for many years. I shoot myself, having married into a family fanatically dedicated to the pursuit of snipe and woodcock and, in addition to our local varied winter shooting programme, our holidays have usually been spent in bogs, heather and swamps wherever these fascinating birds are to be

33. A very good head: Brittas Quail owned by the author.

found. I quickly decided that if I was not to spend the rest of my life in the role of dogsbody I had better join in, so I equipped myself with dung-coloured waterproofed clothing and as a reward for producing our first son I was provided with a 'ladies'' 12-bore and told to learn how to use it.

We originally decided to add an Irish Water Spaniel to our entourage primarily to put up woodcock but we quickly discovered that we had all things to all men. That this is so is underlined by the fact that they run in spaniel trials in the United Kingdom and in retriever trials in Ireland. They hunt the thickest undergrowth, retrieve, being large enough to carry a greylag goose, and not too proud to pick a snipe, though it must be said that if guns regularly shoot jack snipe a protest will probably be lodged.

The bouyant waterproof coat, large feet, long legs and barrel-shaped body are a swimmer's dream; the dog performs with the effortless ease of the otter in fast-flowing water, entirely oblivious of the cold. Its excellent eyesight and acute nose ensures that it rarely fails to retrieve from water and may be forgiven for putting a bird down on landing in order to have a good shake before continuing to bring it to hand.

I find the Irish Water Spaniel soft-mouthed and gentle, with determination and stamina second to none, tiredness is not a word they understand until the day is done. I cannot recall ever having one refuse to work, no matter how hard the going.

For flighting at dusk they are invaluable. The dog usually indicates the direction of an approaching bird long before I hear it and the brown coat is remarkably inconspicuous. For retrieving from a reedbed they have no parallel.

TRAINING

The training of an Irish Water Spaniel is basically the same as that given to any other gundog, remembering that it is more versatile than most, highly intelligent, sensitive and sometimes wilful, and will probably train you rather than the other way round.

Three things are necessary to the would-be trainer—patience, complete control of one's temper and enough time to devote to the task. Without any one of these it is better not to start. Enough time does not mean the odd Saturday afternoon, it means a little time every day and not when dog or trainer are tired or hungry. More will be achieved in ten minutes after breakfast then in sixty minutes before dinner after a hard day's work.

Before any training can begin confidence and trust must be established between dog and trainer and I repeat that I believe that this is most easily achieved by bringing the puppy up in the home—under the kitchen table in ours—where he learns his manners naturally by daily lessons and example, where everything he does can be watched and gently approved or reproved and the basics of 'Sit', 'Stay', 'Come', and that vital command

No are firmly ingrained in the normal round of daily happenings before any formal training begins at all. The puppy becomes familiar and at ease with the family, their voices and the comings and goings of the household. He watches and learns from the behaviour of the older dogs, a very important part of the puppy's education; elderly lady Water Spaniels allow no nonsense from young upstarts and do not mince their words. Indiscretions are promptly corrected and not likely to be repeated. The disdain with which my old lady would consider some youthful misdemeanour has sent many puppies scuttling for the safety of the table legs.

The Irish Water Spaniel learns quickly and wishes to please but he matures slowly and, as with other species, adolescent rebellion has to be got through with patience and understanding. As surely as the student's jeans and long hair change to the city suit when respectability of a profession is upon him, so will the wilful puppy mature into a responsible performer in the field. But it calls for calm nerves. Let the puppy show you the speed at which he wants to progress, making sure that each lesson has been firmly fixed before proceeding to the next.

If the puppy is not one I have bred myself, I like to have it from the breeder at eight weeks old and feed, exercise and teach it myself. We make it a rule in the family, however, that we all use the same words of command. These should be short and distinct one from the other; 'Sit' and 'Seek', for instance, are too alike when spoken quickly. If the puppy picks anything up we have an absolute ban on chasing it or snatching the article away; this is difficult to remember if it happens to be one's only pair of unladdered tights but it is, after all, one's own fault and many sensitive dogs can be put off retrieving altogether by an early experience of an agonised scream and flurry of waving arms from its by now adored and heretofore calm owner, to whom he thinks he is rendering a welcome service. Always congratulate him for bringing anything to you in the early days, replace it and then tell him to leave it there—he will soon learn the things to be avoided. Accidents are bound to happen and you should be prepared for a few casualties during the training of a young dog.

Between six and nine months, but only when the obedience of the puppy is firmly established, more formal training begins. Just exactly how the education proceeds depends on the individual requirements and methods of the trainer and type of game available. Before live birds are encountered, retrieving with a dummy should be taught, and the dog gradually accustomed to the sound of shot. I start with a paper bag and progress to a ·410 fired by an assistant at a distance from the dog. It is useful to introduce the puppy to the chicken run and tame rabbits so that he clearly understands from the start that these things are taboo. *Any* interest in sheep can be scotched by a short encounter with a mature ram.

It should be made clear to the puppy that formal lessons are serious and a regular routine established. The choice of suitable training ground is

34. The Irish Water Spaniel.

important and careful thought beforehand can avoid flagrant disobedience by not providing the opportunity for it to occur. I usually start retrieving, for instance, in a small walled garden which gives complete privacy and where the puppy can concentrate on me alone; if he is slow to come back with the retrieved dummy I can move to the gate as though to leave and the thought of being left behind in a strange place is enough to galvanise the most dilatory puppy.

Keep the lessons short; resist the temptation to try just one more when things are going well. Stop immediately if they go badly and start again the next day with more elementary lessons. The Irish Water Spaniel will not be coerced; encouragement and congratulation should be lavished for good behaviour; disapproval in the voice and eye are enough reproof for disobedience—chastisement is fatal.

The Irish Water Spaniel needs no special training to enter water, keeping him out is much more likely to be the problem. One of mine, when about six months old, jumped into a huge water tank three feet off the ground and we had to have all our tanks covered with netting as she could get in unaided but not out. When taken to the sea she always dashed straight into the water, turned round, sat down and would stay there all day, oblivious of waves breaking over her head, unless we actually restrained her. However, they should be taught to enter the water quietly and also to get in and out of boats—over the stern is easiest and safest. It is also advisable to teach response to hand signals; these are much more use to a dog in the water than a lot of shouted instructions which he probably cannot hear anyway if his ears are full of water.

Don't expect to achieve success in six easy lessons. The individuality of the Irish Water Spaniel is so great that, of course, they have their on and off days, though on reflection I find that if the dog has not performed as well as usual it is through some fault of mine—giving a muddled instruction perhaps or asking it to do something undignified—my late and greatly lamented bitch annually came home from our local charity gymkhana with the first prize in the obedience test until the time I told her to sit beside a Chihúahua barely tall enough to see over the grass—this was too much for Caragh; she walked with great dignity from the ring and nothing would persuade her to return. The fault was mine, not hers.

AVAILABILITY

The Irish Water Spaniel is not the sort of dog you potter out to buy on a Sunday afternoon. They are not found in pet shops or large commercial kennels advertising a vast selection of dogs. It is relatively rare, bred by its devotees who are careful to ensure that its great qualities are maintained. Interest, however, has increased considerably in the past few years and more are to be seen in the field and on the bench than for a long time. Classes are provided at Cruft's and at dog shows at Windsor, Blackpool and Birmingham, and there are usually representatives at the Game Fair and at the Show Spaniel Field Day. Selection of puppies should be made with care and it is advisable to consult the secretaries of the various breed societies, The Irish Water Spaniel Club in Ireland, The Irish Water Spaniel Club of America and, in the United Kingdom, the Irish Water Spaniel Association, whose present Hon. Secretary, Mr N. Waters of 'Zanfi', Drury Lane, Summerwood, Rodington, near Shrewsbury, will provide information regarding puppies available or litters expected and is full of useful advice about currently successful working animals.

In my experience puppies are mostly spoken for before birth and I have waited anxiously with a whelping bitch to discover if a litter was large enough to fulfil all the bookings. However, in choosing a puppy, the temperament is all-important and the pedigree of both parents should be inspected to see that they are not closely related, while it is wise to be satisfied that both parents work and have stable temperaments. They can be shy when puppies and mature slowly, but with careful unhurried training they develop into the most indomitable animals, able to rise to the occasion in moments of crisis, bold and fearless, who stand their ground in the face of any adversity.

An Irish Water Spaniel is a treasure, a well-trained one a treasure beyond price.

8 Welsh Springer Spaniel

Although the Welsh Springer Spaniel is claimed to be of very pure and ancient origin, there are no records available to establish this. The Spaniels depicted in old prints were usually red and white and do bear some resemblance to the present-day Welsh Springer. The fact that they breed consistently to a uniform type does proclaim their ancient lineage and there can be no doubt that they are not a 'manufactured' breed.

At the turn of the century the breed came into prominence when the late Mr A. T. Williams, Ynis-y-Gerwn, Neath, in the Vale of Glamorgan, appeared with his Welsh Springer Spaniels at field trials and shows. Mr Williams claimed that these dogs had been bred and used for sporting purposes by his family for nearly a century.

The breed seems to have been confined to Glamorganshire and a number of landed gentlemen in that County had kennels of these Spaniels, used entirely for shooting, including Sir John Talbot Llewellyn of Penllergare, Colonel Lewis of Greenmeadow, near Cardiff, the Rev. Horatio Thomas of Pentyrch and Colonel J. Blandy-Jenkins of Llanharan. The latter gentleman's Spaniels had been at Llanharan at that time as long as anyone could remember, and were well-known as the Llanharan strain.

Up to the time of the 1914/18 war breeding was carried out with care in order to produce the very best working dogs, and as far as we know no alien blood was permitted, the strains being kept entirely pure. There is a possibility that the Breton, or Brittany Spaniel may have been introduced as this is a very similar type of Spaniel to the Welsh Springer and it is likely that dogs of this breed did come into Glamorganshire on the trading vessels which visited the South Wales Ports. There is, of course, no evidence to support this but the writer has seen in France, Brittany Spaniels which very closely resemble the Welsh Springer Spaniel indeed.

Puppies were often gifts to shooting friends and as a result passed into different hands so that the breed gradually became distributed throughout Glamorganshire and eventually farther afield.

Field Trials for Spaniels started in 1899 and the Sporting Spaniel Society, which existed at that time, held its third Field Trial Meeting in 1900 at Ynis-y-Gerwn, Mr A. T. Williams' Estate. Mr Williams' team of

Welsh Springers, amusingly named Rush, Brush, Flush and Dash, all of Gerwn, appeared for the first time and carried all before them, beating three other teams of other varieties of Spaniels. It was thought that this success was due to the fact that the dogs were working on familiar ground, but in 1901 Mr Williams' team was again successful when the Trials were held at Sutton Scarsdale. Mr Williams' Spaniels continued to be all-conquering, and in 1904, with a different team, Rant, Belinda and Roger of Gerwn, won the Sporting Spaniel Club's Team Stake at Sir Watkin Williams Wynn's well-known place, Wynnstay. For good measure he also took first and second places in the Brace Stake, with Roger and Bella of Gerwn, and Glory and Flo of Gerwn respectively.

Mr A. T. Williams was responsible for the breed being recognised as a separate variety by the Kennel Club. In 1895 Colonel John Blandy-Jenkins, J.P., Chairman of the Glamorganshire County Council, gave to Mr Williams a very fine Spaniel called Flint and the notes in my possession, written by Mr Williams, say he was the last Spaniel bred or kept at Llanharan House. Flint was given to him as a great personal kindness on condition that Mr Williams would never part with him. The dog was registered at the Kennel Club as Corrin and as the Kennel Club did not then recognise a separate breed known as Welsh Springer Spaniels, he was registered as a Cocker! His owner said of him that he was the best looking sporting Spaniel in England. During the years 1899 to 1901 inclusive he was exhibited at a number of Shows with remarkable success. In classes for sporting Spaniels, which at that time included any variety of Spaniel, he was unbeaten. Mr Williams said that he was a true representative of the Llanharan family of Spaniels and in addition to his good looks, he was a very remarkable worker. The best, said Mr Williams, that he had

35. A group of Welsh Springers . . . perhaps the most attractive of the Springer tribe.
Peter Moxon

ever had. It was in 1902 that the Kennel Club admitted Welsh Springers as a separate breed on their Register and Corrin's registration was accordingly altered.

Corrin's sire was Dash and his dam, Busy, both unregistered, but it is believed both were red and white. Corrin probably served a great number of bitches, many of which were not red and white. Corrin was mated to Mena of Gerwn and in 1903 produced the great Champion Rover of Gerwn, said to be the best Welsh Springer ever produced in the Gerwn Kennels. Mated in 1910 to a red and white bitch, Belle of Gerwn, pedigree unknown, Rover sired two great Spaniels, Duke of Gerwn, a black and white, and Roverson of Gerwn, a liver and white! Both were registered as English Springers. Duke was placed third in the first Spaniel Championship in 1914. The pedigrees of nearly all the English Springer Field Trial winners prior to the last war traced back to Roverson.

At the beginning of the 20th Century there was certainly great confusion in the registration of Spaniels at the Kennel Club—a larger dog was registered as a Springer, a smaller one as a Cocker; a red and white one was a Welsh Springer, a liver and white or black and white, an English Springer. Corrin's son, Guy of Gerwn, was registered as an English Springer. Very many of the present day field trial English Springer's pedigrees can be traced back to the Welsh Springers at the beginning of the century. Those Welsh Springers left their mark as can easily be seen by the great similarity between the present day field trial English Springer and Welsh Springer. Apart from colour, many a field trial English Springer would make a very good Welsh Springer show dog. Both have the same small vine-leaf-shaped ear and a similar distinctive head, entirely different from the modern show English Springer's long pendulous ear and exaggerated square flews. As

36. A strong workmanlike 'Welshman', able to go all day.
Peter Moxon

well as looks, how much of the brilliant working ability of the top class field trial English Springer of the present day is inherited from the early Welsh Springer?

Mr A. T. Williams exported quite a few of his dogs to Mr Hobart Ames of Connecticut in the U.S.A. who used these dogs for shooting there. Unfortunately, the untimely death of Mr A. T. Williams just before the 1914/18 war, and the war itself, caused a very serious setback to the breed. No breeding took place during the war and the owners of the larger kennnels, forseeing the difficulties of feeding, had most of their dogs put down. The Welsh Spaniel Club which had been formed to foster the breed, ceased to exist.

After the war a number of people again started to breed, but the large kennels of shooting dogs had now all gone. Nevertheless, there were a lot of people with a few dogs, and in 1922 the late Colonel J. Downes-Powell called a meeting of interested people at Cardiff and the new Welsh Springer Spaniel Club was created in 1922, and was registered by the Kennel Club in January 1923. By the end of that year it had nearly 100 members and the first field trials were held by the new Club on Lord Tredegar's ground at Ruperra Park. A Stake was held for Welsh Springers which brought out six entries. There was another for Cockers with eight entries and an Any Variety Stake which had five or six entries. The best Welsh Springer at that time was Bacchus. He not only won the Welsh Springer Spaniel Stake but won the Open Stake for Any Variety Spaniel at the Northern & Midlands Field Trial Society's Meeting, and also went on to obtain a Diploma of Merit in the Spaniel Championship.

The trials in 1923 were a great success and the following year the Club's new President, Captain A. Talbot Fletcher, invited the Club to go to his Estate at Margam, near Port Talbot. There a two day Meeting was held and there were eight Stakes. The trials were most successful and the Welsh Springer Spaniel Club's name was made as a very successsful field trail organisation.

The Club continued field trials annually up to 1938 and after that trials ceased during the second world war.

During the two wars several Welsh Springers distinguished themselves. In addition to Bacchus, Tess of Shill owned by Mrs S. Horsfield, gained a Diploma of Merit in the Spaniel Championship in 1925. Colonel and Mrs Horsfield won eight Stakes with Welsh Springers. Colonel N. W. Tredinnick's Ginger Lad, and Mr R. James Shot of Cannonmoor each won four Stakes. Mr R. H. Sprake's Legacy Lex won the Welsh Springer Club Open Stake in 1931 and then went on to win the Eastern Counties Spaniel Club's Open Stake for Any Variety Spaniel. In 1929 Lt. Col. J. Downes-Powell brought out Marksman O'Mathern which won the Novice and Open Stakes for Welsh Springers. This dog was regarded as one of the most perfect specimens of a Welsh Springer of all time. He was by Champion Merglan Bang out of Goitre Lass and had the distinctive flesh-

37. Welsh Springers are fearless in water.
Peter Moxon

coloured nostrils and hazel eyes. Although his owner had sold many Welsh Springers to the United States, it is understood that he received an offer of no less than £2,500 for this dog, a very large sum indeed in those days, but Marksman remained his Master's shooting companion for his lifetime.

After the cessation of hostilities, the Club ran a one day field trial at Margam Park in 1946, there being one Open Stake for Any Variety Spaniel. One Welsh Springer only was entered—Rockstead Footstrew owned by Mr Ronald MacDonald. In the following years the Club's trials came back to normal and in addition to Stakes for Any Variety Spaniels, a Novice and Open Stake were run each year for Welsh Springers which were fairly well supported. Several well-known professional handlers trained and handled Welsh Springers for their owners such as the late John Forbes, John McQueen, Ronald MacDonald, and several others. With one or two exceptions, I regret to say the standard of Welsh Springers did not compare with the English Springer. Several Welshmen ran in any variety stakes but no major awards were won and since the end of the War no Welsh Springer has qualified for the Spaniel Championship.

Rushbrooke Racer was the best Welsh Springer and created a record by winning the Open Stake for Welsh Springers for four years in succession, 1949 to 1952.

The separate Stakes for Welsh Springers were continued by the Club up to 1957 but were then discontinued because support was not forthcoming. In 1960 an attempt was made to renew interest in Stakes for Welsh Springers only and a Meeting was held at Compton Abdale in Gloucestershire and was quite well supported. Apart from the winning dogs the standard of training was very low. The Stake was won by Mrs D. Morriss' Stokecourt Susan. Meetings fixed in subsequent years had to be abandoned because of lack of entries. For several years Mrs D. Morriss competed in Any Variety Spaniel Stakes with Stokecourt Sam, which had a number of awards, but regrettably no Welsh Springers are now seen at field trials. As far as the writer is aware there are no Welsh Springers now being trained to field trial standards.

That this regrettable state of affairs should have come about is a tragedy which has happened not only to Welsh Springers but to other breeds of Spaniels such as the Field and Clumber. The number of Welsh Springers now being bred and registered at the Kennel Club far exceeds anything imagined, even 25 years ago, but alas their enthusiastic owners are concerned only with the show bench and, generally speaking, working qualities are ignored by breeders.

The Welsh Springer is a very beautiful looking Spaniel and an ideal helpmate for the rough shooter. What opportunities there are for shooting men to seek out the best working Welsh Springers they can find and breed once again for working qualities! It could be done, as here and there over the country are shooting men who still own and shoot over Welsh Springers which never come into the public eye.

DESCRIPTION

I shall always remember the year 1928 and the first Welsh Springer Spaniel I ever had. He was given to me as a present by my mother—I was still a boy and came home to find a young dog of 18 months waiting for me, and I gazed at him in admiration. I thought he was the most beautiful Spaniel I had ever seen. I have always regarded the Welsh Springer Spaniel as the most attractive looking of the Spaniel tribe. Unfortunately the dog's pedigree had been lost but I took him to the old gamekeeper in our village who immediately pronounced him a perfect specimen of the Llanharan Spaniel. He wrote a Certificate in his long spidery handwriting accordingly. I sent it to the Kennel Club and registered the dog as a Welsh Springer Spaniel 'Pedigree unknown.'

Only one colour is permitted, a dark rich red and white—a real red on white, which in a Welsh Springer always seem to me to have a silvery sheen like virgin snow, except of course when the dog has been working and is wet and muddy. This distinctive colouring serves well in the shooting field to indicate the dog's whereabouts but there have been odd occasions when the dog's colouring has been a perfect camouflage when hunting in dead bracken, with here and there remains of melting snow upon it.

38. Accepting a retrieve.
Peter Moxon

In size the breed occupies a position midway between the smaller Cocker and the larger English Springer. Although the standard of the breed published by the Welsh Springer Spaniel Club in 1923 does not state a height at shoulder, I found that most of my own have been 17½ in, and 18½ in. respectively for dogs and bitches. It is interesting to note that the great Corrin weighed 47 lb. He was a very heavy dog.

The standard of the breed drawn up by the Club describes it as 'a symmetrical, compact, strong, very active dog, not stilty, obviously built for endurance and hard work. A quick and active mover displaying plenty of push and drive.'

How better can a working Spaniel be described? He must be strong and muscular, well boned with a deep brisket having plenty of heart room; powerful hindquarters providing the driving power, great forelegs and altogether length of body proportionate to the length of leg, so that the whole appearance is of a very well balanced dog which is finally set off with a well set on tail carried low, very lively in action. A dog of this description is capable of working all day on any sort of ground in any sort of weather.

A distinctive feature of the breed is its very small ear, not long and pendulous, but gradually narrowing towards the tip and shaped somewhat like a vine leaf.

The tail is docked but at least half its length should be left and the docking should not be the removal of the larger part of the tail, as is done to the dogs one sees on the Show Bench of today. Their whole appearance is spoilt by the stubby stump left to them devoid of all feather. The tail is the dog's means of expression and its lively action tells the shooter what is afoot.

Originally the nostrils were always flesh-coloured and the eyes hazel as would be expected from the dog's colouring. Nowadays, however, the eyes are usually dark and the nostrils black which does make a very handsome dog indeed, but the flesh-coloured nose and hazel eyes constantly reappear. In the 1920s there was great controversy when the black nose and dark eyes first appeared. About 1925 a Mr. Lewis of Talybont-on-Usk in Breconshire, had a very good bitch known as Goitre Lass. It is believed that she had some English Springer blood in her veins but I think this unlikely as the liver and white English Springer almost invariably has a flesh-coloured nose. It is more probable that this bitch had some Cocker blood and indeed there were suggestions at the time that a black Cocker had been used to produce the black nose and dark eyes.

At any rate Goitre Lass produced a large number of puppies in her lifetime through regular matings with a very fine dog, Champion Merglan Bang. It is believed she produced a litter every nine months for a number of years and certainly all were the correct red and white colour, but she did produce a large number of puppies with darker noses and eyes. By the early 1930s the black nose and dark eye was becoming exceedingly

39. The author's Rushbrooke Racer. A great winner in the '50s.

popular and gradually the flesh-coloured nose and hazel eye was seen less and less at Shows. So many puppies were bred from this mating, that it is probably difficult to find any Welsh Springer alive today which has not in its veins, the blood of the Champion Merglam Bang/Goitre Lass combination.

Personally, I still prefer the hazel eye, not the light gooseberry eye, and the flesh-coloured nose, and I have always found that my best workers had this colouring.

WORKING ABILITY

As a worker a really good Welsh Springer is ideal for the man who loves shooting over Spaniels, but the good ones are not very easy to find these days. There are no kennels breeding Welsh Springers exclusively for work and the continuous breeding by Show enthusiasts of Welsh Springers, just as in the other breeds of gundog, has been to its detriment as a worker.

A good Welsh Springer possesses tremendous drive and dash and has an amazing nose. He is a natural line hunter. This great attribute has been something of a drawback for the breed, particularly in training dogs for field trials. Unless well-trained the Welsh Springer is inclined to follow foot scents instead of ignoring these and quartering his ground, ever seeking the squatting game on his beat. He is at his best in thick cover and he will face the toughest quite fearlessly. His line hunting ability does make him exceptionally good in the retrieving of runners.

This can be illustrated by the story of the young Welsh Springer bitch, Rushbrooke Runner, competing at the first field trial immediately after the war in 1947 on a Welsh mountainside in Breconshire. It was her first season and the young bitch did not really understand what it was all about. She had competed in the Novice Stake for Welsh Springers without distinguishing herself and she was now competing in the Open. In a small copse high up on the hill she flushed a hen pheasant which flew out over the trees and away down the hill. She sat and watched its going and a shot was heard. A minute or two later the Judge called handler and dog out of the wood and pointing down the open grassy hillside, said: 'The bird fell down there, it's only lightly winged tipped; see if your bitch can get it'. The bitch was sent out. She went at once where indicated and struck the fall. Down went her nose and without any further command she went down the hillside at full gallop, to a cleft in the hillside where a stream went tumbling down to the valley beneath. At the stream she stopped, crossed it, cast up hill for 30 yards or so, but had lost the line. Without hesitation she turned and galloped back to the spot where she had crossed, picked up the line again and went down the hill by the side of the stream as hard as she could go. Her handler could do nothing except watch, and indeed by this time she was well over 300 yards away. Continuing going, she came to a spot where the ground beside the stream was covered in rushes. Into this she disappeared.

From the vantage point high upon the hill the pheasant was seen to run out from the opposite side of the rushes with the Spaniel close behind. In a moment she had caught the bird and without hesitation turned and came galloping straight back up the hill with the hen pheasant, with its head up, eyes bright, held firmly but softly. She delivered it gently to her handler. It was the first running pheasant she had ever retrieved. All her work had previously been on rabbits. She had won the Open Stake. This illustrates not only the Welsh Springer's extremely good nose, but also its retrieving ability and initiative.

I had often heard it said that the Welsh Springer Spaniel is not a good retriever, but generally I found those that I had very good. There were some, of course, as in all breeds which were not interested in retrieving, but if they are bred right, then I think they are exceptional.

There were many Welsh Springers which were never called upon to retrieve and probably these dogs were bred from without any thought being given to whether the dogs were natural retrievers or not. It is known that Colonel Lewis of Greenmeadow hunted his Welsh Springers in teams on the bracken-covered Welsh hillsides. These dogs were particularly well trained, hunted at tremendous speed and with the greatest enthusiasm, driving rabbits forward to waiting guns on rides which were cut at regular intervals through the bracken. On coming to a ride the dogs all dropped. The guns then moved on to the next ride whilst the picking up was done by the Colonel's Retrievers. The Spaniels were not used for this at all. The late Lt. Col. Downes-Powell told me that once he had attended one of these shoots and saw one of the Welsh Springers, instead of dropping when coming to the ride, go forward and retrieve a rabbit. Colonel Lewis shouted to his keeper 'J — shoot that b . . . dog. What do you think I keep Retrievers for?' Whereupon Colonel Downes-Powell immediately said 'If you do not want that Spaniel will you let me have it?' He was presented with the dog there and then and Colonel Downes-Powell later won with it at field trials.

That Welsh Springer was a natural retriever, but there may have been many, never used to retrieve, which were not, and the offspring of these were likely non-retrievers. In the early field trials often only one dog in a working team was required to retrieve. The main purpose of the dogs was to hunt and find game.

Today, if a shooting man can find a good Welsh Springer, it is a pearl beyond price, but to find such a dog will not be easy. Having found a puppy which shows all the promise of making the sort of shooting dog I have described, the owner will need all the skill he has to train him with care and affection to produce the perfect shooting companion. I say care advisedly because a Welsh Springer needs careful handling. He is resentful of harsh treatment. The trainer who ill-treats his Welsh Springer will have no success whatsoever in training him, but for the man he loves, the Welsh Springer will give of his best.

CARE AND MAINTENANCE

So much has been written about the care and maintenance of gundogs that it is superfluous for me to say anything about this, because the Welsh Springer requires no different care or feeding from any other working Spaniel. Naturally he wants plenty of exercise as a working dog, and I think the one-dog owner cannot do better than let his young Springer be his constant companion as far as this is practicable. Let him come into the house and be about and meet people. Let him play with children during his youth. A dog kept alone in a kennel cannot be a happy dog, and a dog which is not a happy dog, is not going to give of his best.

If he has companions, then let him live in the kennels by all means, but even then, have him out with you on all possible occasions. Let him know everything that is going on because you want a bold dog. A nervous frightened dog is no use to anyone.

For many years I kept a number of Welsh Springers and I always kept them together in a large wired-in run in an orchard. I found they did very well in these surroundings and I am sure that well drained soil is far better for a dog, than spending the whole of its day on cold concrete. At night let him have his wooden bench well off the ground, free from draughts.

As far as feeding is concerned, the Welsh Springer seems to thrive on almost anything, as most dogs seem to do. A large proportion of my dogs' diet was raw meat, usually raw tripe. They never had any cooked meat, all did very well and skin disease was almost unknown. The Welsh Springer is very hardy, very adaptable and requires no special attention.

9 Clumber Spaniel

The origin of the Clumber Spaniel is shrouded in mystery and their ancestry is a matter of mere conjecture. However, it is certain that they derive their name from Clumber Park, one time Nottinghamshire seat of the Dukes of Newcastle. They made their first appearance in this country at the end of the 18th Century and one theory is that they came from France as a gift from the Duc de Noailles, for he presented his entire Kennel to the second Duke of Newcastle. In my opinion, it is most unlikely that this is the true story, as there are no spaniels in France at the present time which even remotely resemble the Clumber nor, as far as I can ascertain, are there any records or mention of such a breed. However, it is possible that this was a local strain bred by the Duc and no dogs were ever sold and all were retained on the estate. Of course, it must be remembered that many early records of sporting dogs were lost during the French Revolution.

Another suggestion, is that the breed was evolved by the crossing of the French Bassett with the old English Spaniel, but again this is unlikely as early pictures of the breed show no trace of such a cross except perhaps in the eye, as some Clumbers show definite haw as does the Bassett. If a hound cross had been used one would expect the resulting progeny to speak on a line, but the Clumber is a mute hunter.

One of the best early pictures of the Clumber was one which used to hang in the Clumber House and was painted by Francis Wheatly R.A., in 1788; it depicts the second Duke of Newcastle mounted on his shooting pony with Colonel Litchfield also mounted and Mansell, the headkeeper, and a team of spaniels of undeniable Clumber type.

Mr James Farrow who was one of the great experts in the breed at the latter part of the last Century, was of the opinion that the most likely origin of the Clumber was from the old Sporting Blenheim Spaniel of many centuries ago. This dog was red and white, and lemon and white in colour and had little resemblance to the tiny Blenheims which are found today. Old pictures of the Blenheims show a marked resemblance to the dogs shown in Wheatly's picture. It is interesting to note that in all the early pictures of Clumber Spaniels they were much more heavily marked

40. A Clumber retrieving to hand. *Miss A. Roslin-Williams*

41. Lynchine Lynette, a 7-month-old Clumber retrieving a wood-pigeon in good style.

than are the present day specimens.

When the breed was first introduced to this country, it remained in the hands of a comparatively small number of people; practically all the leading Kennels were to be found in the 'Dukeries' area of Nottinghamshire, and large studs were maintained by the Duke of Newcastle at Clumber Park, the Duke of Portland at Welbeck Abbey, Mr Foljambe at Osberton Hall, and the Earl Manvers at Thoresby Park. The Duke of Devonshire also had a strong Kennel at Hardwick Hall in Derbyshire. At this period the breed was held in high esteem as a sporting spaniel and Mr Foljambe of Osberton Hall recently showed me a most interesting catalogue of an auction of Clumbers by Messrs Tattersall and Son of Hyde Park in the year 1851, being the dispersal sale of the deceased Duke of Newcastle, and comprising a dozen trained dogs and ten youngsters. The highest price paid was 64 guineas for the two-years-old bitch Busy and a litter of five, whelped on July 18th, 1850, made a total of 107 guineas.

Some fifty years later in January, 1899, the first field trials were held at Sutton Scarsdale, the home of Mr William Arkwright, who maintained a strong Kennel of Clumbers in addition to his more famous Pointers. These trials were held under the auspices of the Sporting Spaniel Society and it is interesting to observe that at the early trials the Clumber was almost invincible, especially that truly great bitch Beechgrove Bee owned by Mr Winton Smith. The following list of awards shows how great were the Clumbers at that time: Sporting Spaniel Society, January 1899, second All Aged, second Puppy; December 1899, First All Aged, First Puppy, Second Brace and Championship; 1900 Second Puppy, First All Aged and Championship; 1901 Second Winners; 1903 First Non Winners, First Winners and Championship; Spaniel Club 1899, First and Second Open, First and Second Retrieving Spaniels, First Brace and Championships; 1900 First Open, First and Second Retrieving Spaniels, First Brace and

Championships; 1901 Second Open, Second Novice, Second Maiden, Second Brace; 1902 First and Second Maiden, First and Second Novice, First Brace; 1903 First and Third Maiden, Third Novice, First Open, Second Brace and Championships; 1904 Second Non Winners, Second Open, First Brace, Second Team; 1905 Second Non Winners, Second and Third Open.

During this period the breed was at its zenith as a working dog and I do not suppose such a record will ever again be realised.

The popularity of the Clumber was further strengthened when, in 1925, the late King George V re-established the breed at Sandringham, where they were used as a team for covert shooting on the Royal Estate. His Majesty held them in high esteem as they were eminently suitable for flushing game from the rhododendrons which abound at Sandringham. The Clumber is a heavy dog and having a placid nature was ideal for this type of work.

What is the position of the working Clumber today? Alas, not a single Clumber is to be found on the great estates of Nottinghamshire and it is only seen in relatively small numbers in the shooting field.

The cause of this decline is difficult to understand, but I suppose the truly magnificent appearance of the breed has contributed to its downfall as a workman, for between the wars it was largely in the hands of people primarily interested in exhibition; insufficient emphasis was placed on the working side, and head properties and exaggerated size and bone were the main objectives. Happily, this state of affairs has now been rectified and there are many people breeding and showing a truly workmanlike type of dog.

One should remember that without the dog show the breed would most likely have become extinct and we should therefore be grateful to that small band of people who kept the breed going when it was out of fashion.

Two of the most noteable supporters during this difficult era were Mr R. Cape of Egham who added to his Kennel by acquiring the majority of the Sandringham Clumbers on the death of His Majesty King George V and the late Miss Read of Corbridge who maintained a strong Kennel of show and working Clumbers throughout her long life.

At this stage I think it would be prudent to describe the correct type of dog which breeders should aim to produce. The appearance of the Clumber should be that of a heavy massive dog with a thoughtful expression. A dog should weigh around 60 to 70 pounds, bitches 45 to 60 pounds; one can take a broad view of these figures so long as the animal is in good proportions. Although he is a heavy dog he must be sound and very active, as his job is to flush the game from thick cover.

The great feature of the Clumber is his head. Quite different from any other spaniel breeds, it is large, square and massive, broad on top, with heavy brows and a deep stop, heavy muzzle with well developed flews, square flesh-coloured nose, dark amber eyes, slightly sunk, large vine shaped ears, hanging slightly forward, feathered but not below the leathers. His body is long and heavy, near the ground with well sprung ribs, back straight, neck fairly long and thick. He must have powerful hind action, which necessitates a good turn of stifle and good angulation of the hocks to give him driving power.

His colour is mainly white with lemon or orange markings, lemon prefered; large patches of colour on the body are not desirable. Another feature of the Clumber is the fairly long dock which should be well feathered.

The Clumber is now making quite a comeback in the show ring and nowadays many first-class specimens are to be seen at all our leading Shows. Generally speaking classes are well filled, and the dogs have greatly improved in movement. A few years ago there were many more bad movers than sound dogs. Hindquarters were straight with no turn of stifle and as a result drive was lacking; fronts were also far from straight. But now many really good dogs are on view and bad movement is the exception; the dogs which are winning today look capable of doing an honest day's work.

I am sorry to say that on the working side similar strides have not been made and there are no Clumbers running in present day field trials, although more dogs are to be seen every year in the shooting field.

It must be realised that this breed is quite different from all other spaniels, for it is relatively slow and not spectacular in its work, but it nevertheless has many virtues as a worker.

I think I can say that all Clumbers have excellent noses and are very persevering in their work. Sometimes lack of speed can be an asset, expecially on a bad scenting day. There is an old saying that a hound is only as fast as his nose, and this also applies to a gundog. All the Clumbers I have trained have had exceptionally good mouths, have been natural water dogs and have usually faced the most punishing cover, reminding one of a miniature tank in action. I am sure that there is a great future for the breed providing they are used for purposes for which they have natural talent.

A team of Clumbers used instead of beaters for covert shooting would be more efficient and far less expensive than five times the number of human beings. In this type of work their slow and methodical mode of working would be a distinct advantage.

Due to their tranquil disposition the Clumber makes an excellent non-slip retriever as his temperament is more akin to that of a retriever than the spaniel breeds. Perhaps his best role is that of a rough-shooter's dog and he is in his element pushing out the old cock pheasant from thick hedgerows and bull-dozing rabbits out of virtually impenetrable bramble patches. There are many people shooting today who, to say the least, have passed the first flush of youth and would find the plodding pace of

42. A Clumber working on a shoot. Mr Sanders with his Frastan Envoy.

the Clumber greatly to their liking.

The training is similar to that of any of the spaniel breeds and many abler pens than mine have previously dealt with this subject in great detail. However, I do not consider that Clumbers are the easiest of spaniels to train as they have a rather independent nature and much care and time must be spent in getting them really steady on the drop. This, in my opinion, is the most important lesson to be learnt; all the subsequent training is dependent on a high degree of efficiency being attained at this stage. One must not be satisfied until the dog will drop instantly to command, shot and whistle.

It must be remembered that when once introduced to game it is much more difficult to maintain the high state of obedience which we hope has been attained in the preliminary work with dummies. So many shooting people labour under the mishapprehension that a dog becomes steadier with age, whereas, in actual fact, unless a dog is rock-steady in his first season he will gradually become wilder as he gets older.

I have also found that many Clumbers are at first disinclined to retrieve natural game, although they retrieve the dummy in exemplary fashion.

It is, therefore, most important that the first bird the dog is asked to retrieve should be stone-cold and the wings tied down to the body. Only tight feathered birds should be chosen for the first few retrieves and never on any account be tempted to use a woodpigeon as the feathers are so loose that the dog is likely to drop the bird to enable it to rid its mouth of the feathers which it finds so distasteful.

In my experience I have found it advantageous completely to cover the dummy with wings of game birds during the later stages of training. When the dog will retrieve a cold bird readily, a freshly killed bird may be used. In the early training days lessons should be of very short duration, say five or ten minutes at a time, but repeated at very frequent intervals two or three times a day. If this method is adopted the pupil does not become bored, but at the same time the frequent repetition stamps the lesson firmly in his mind.

This advice may seem very elementary but in the case of the Clumber I consider it most important and, of course, when introduced to the shooting field be absolutely certain that the first few birds retrieved are dead; if there are runners to be collected send an older dog.

I know that it is very tempting to send out a promising dog for a runner, but I am certain it pays to make haste slowly, as so many dogs are ruined during their first few days of actual shooting. One should always remember that a wild and useless dog is a result of bad training rather than any inherent failing in the animal. A gay, well disciplined Clumber is a true friend and a delightful shooting companion.

Clumbers are more readily available now than they were a few years ago and puppies are advertised from time to time in the dog and sporting press. If any difficulty is experienced in acquiring a puppy I suggest that an enquiry to the Secretary of the Clumber Spaniel Club would in all probability solve the problem. This Club, which was founded in 1904, is the Society which looks after the interests of the breed and has a membership of about a hundred including many people from overseas.

Clumber puppies are delightful bundles of white fur bearing some resemblance to young polar bears. The rearing of the puppies is a very important factor as one wants substance and the characteristic heaviness, but at the same time they must be kept active and not over-weight or one soon has the puppy unsound. They want good quality nourishing food as opposed to great quantity. I give a morning feed of milk, bone meal and cereal, with $\frac{1}{2}$ lb. raw minced meat and two fish oil capsules at midday, followed by an evening meal of soaked puppy meal and $\frac{1}{2}$ lb. cooked meat—this is for puppies up to about three months. I increase the meat to $1\frac{1}{2}$ lb. a day as they get older and, of course, the occasional egg, say three times a week, is also very beneficial; a few dry biscuits to chew during the morning are also appreciated.

The Clumber has had many changes in fortune during the two hundred years it has been in this country, but there has always been a nucleus of

43. A close-up of Mr Sanders' Frastan Envoy.

serious breeders who have succeeded in maintaining a true type of dog remarkably free from congenital abnormalities, and such is the case today.

I hope most fervently that it will never become 'fashionable' and be bred in large numbers by indiscriminate breeders, with the consequent evils, but that it will forever remain a noble sportsman as befits its ancient and aristocratic ancestry.

10 Sussex Spaniel

I don't intend to spend a lot of time and space on the history of the breed. This has been so well told by Mr C. A. Phillips and Colonel Claude Caim in their book on spaniels. I would like to say, however, how grateful I am for the privilege of having known both of them. It is so much more profitable to discuss a dog if you have the example in front of you. I remember asking Colonel Caim if he thought my dog too light in the eye and his reply, 'Certainly not. He would look very odd with a dark eye as it would not harmonise with his coat colour.' At once I learnt that the colour of the eye meant quite a lot to him.

From there we went on to discuss coat texture and he told me that the Sussex coat must be like sealskin and that was one of his great priorities in the breed, one to which I have always tried to adhere. I hate a harsh coat or a straggly coat and I find they very often go together. The other observation that I have made over the years, is that more often than not where you have the right texture of coat you get the right colour. I once heard this true colour described as being the colour of wet sand. There should not be any shade of red in the perfect colour, just a shading of gold at the tips. We had great difficulty in finding the right way to describe the colour when we were re-wording the descriptive particulars in the 1930s and finally decided on 'a golden liver with the gold predominating'.

I have met men in their nineties who remember the Sussex spaniel as he was in their youth when every cottager owned one, and they were all big dogs. It was not until the craze for showing took over that we began to see the short-legged elongated specimens. Even then there were long-legged dogs in every litter but they were all turned over to work and only the shorter-legged ones kept for show. When we reviewed the points, we raised the height standard, as we had had so many complaints that the breed was too like toys and no use for the gun. The spirit was still there, but alas the legs were not. Sussex spaniels have always been easy to train and if you can train them yourself you will find how anxious they always are to do everything to please you.

The Sussex head is a smaller edition of the Clumber spaniel, though when I started in the breed the Harviestoun Kennel was still turning out some beautiful specimens and these had a slightly more bloodhound head. I used both in my breeding programme and so we can still get samples of both types. I think it shows what very tough dogs they must have been when you realise that at the end of the last war there were only eight left in the world.

At least there was one member of the famous Wigg Kennel still going in Kent, but it was a very severe tragedy for the breed when he turned out to be quite useless at stud. It would have been such a help if we could have had some different blood from the Hornshill strain. However, it was not to be! I had to get on as best I could with my eight. If they had not, as I say, been a very robust strain I don't know what would have become of us. I thank goodness they have always retained their ability for work. Their noses are said to be better even than the English Springer.

During the last few years, I have tried to revive the habit of getting our champions qualified for work. When I first started breeding I was lucky to have Mr T. J. Greatorex within easy reach and for years he qualified all my dogs for me. You will find that he still has in his heart a very soft spot for the Sussex. When we moved to Suffolk, it was much more difficult as we knew no one who trained spaniels, so quite a few dogs that could otherwise have been qualified were lost to us. Now we have James Wylie's expert help; and we were lucky to get him to judge our first official test at which we actually had eight dogs competing.

One of the characteristics which seems to be disappearing is the old Sussex 'roll' when moving. This is probably due to the fact that very few breeders of the present day have time really to exercise their dogs. They are a very slow maturing breed and I find present-day breeders always in

44. Uncas of Fourclovers. This dog will retrieve anything from anywhere, land or water.
P. R. Lange

such a desperate hurry to show their stock. They just cannot wait until there is something to show. My advice to newcomers is always, in the case of bitches, to let them have three litters while they are young and then you will have plenty of time for showing. Later they will be much better specimens of the breed and, may I add, much better brood bitches when young. Nearly all the bad brood bitches have been left until they were far too old before they were bred from and so much more inclined to be crabby.

Looking back over the years I realise how much luck I must have had in my breeding operations. To breed so early a dog of the calibre of Ch. Okimat was truly astonishing. Really expert gundog men told me he was the best Sussex spaniel they had ever seen and as he remained unbeaten in eight years it would seem to prove this to be true. I certainly have never seen one since to come anywhere near him. Not only was he perfect to look at, apart from a slight wave in his coat, but he had the perfect temperament and was a grand gundog. Mr Greatorex was very disappointed when I would not let him have him for field trials. His half-brother Montezuma, owned by Colonel Downes-Powell, would, according to his owner, have been a field trial champion—he had already won top honours at at least one trial—if the war had not come along to spoil his chances.

Oddly enough Okimat never produced his equal, but Tamenund, one of his sons, was the dog that made the most impression at stud. This dog had 32 bitches to him of which only one failed to breed. He was the only dog I have ever owned to win best in show and this he achieved at the big Liverpool show. He also won top award at one of the lesser shows. The win with most prestige attached was the International Brace class at Cruft's against very hot opposition. This was with my first Sussex Ch. Brosse and his son Ch. The Sagamore. When I bought Brosse from Mr Stephenson Clark I had no thought of showing, but Mr Clark insisted that he was the best of the litter, that his sisters had all won and that I really must show him. So having consulted Mr George Wallwork who advised the same, I duly took him to Cruft's and won Best of Breed. Mr Clark was not interested in showing but frequently ran one in trials of which Broadhurst Bexie was the best. All the men who competed against me at shows in the early days shot over their dogs and it is a great grief to me that today so little interest is taken in the working side.

Of course we had setbacks. In addition to the war, we had eight years in which I recorded total breeding failure. Bitches appeared to be in whelp and then failed to go through with it. However, thanks to Mr Orton of the Sedora Kennels who told me to try Canalin H injections, we cured this, though my vet was extremely sceptical about the whole business and still says that he does not understand why it should work. He has, however, to admit that it does and is even using it with cats.

Another dog which should be mentioned is Sh. Ch. Hiawatha, owned by Miss Dodson of the Weiden Kennels, without him we should not have got very far with our breeding operations. I was lucky to get a bitch by him, out of Muskoka of Fourclovers and she started me off again after eight years of stagnation. During this period I found the breed had shrunk in size and had developed far too much coat. However, we now have five generations from her and size has increased and, thank goodness, we now have more of the real sealskin coat. This is a great joy to me as I have always been very concerned in case we might lose it. There is no other coat in the least like it and if it were to go we could not get it back. This was the chief reason why I viewed the idea of an outcross with apprehension. All I could see was the beautiful coat fading out and a long straggly coat taking its place.

I'm sure the Sussex of the future will owe a great deal to Uncas of Fourclovers. Uncas may have been the last of the Mohicans but he is certainly the first to produce some measure of uniformity into the breed. Although far too big for the show ring, he has all the essentials. A beautiful head, correct coat and colour and, more important, a beautiful temperament. In addition, he is an excellent gundog.

I have just heard that his son Shipden Ambassador becomes the first Sussex for about forty years to qualify and become a full Champion. The last one, I believe, was Miss Wigg's Hornshill Miranda. Uncas's sire Montour, I have always considered to be the best Sussex dog that I have had since Ch. Okimat. It was he that brought back into the breed the long neck which up till then had virtually disappeared. In fact, our supporters in America went so far as to say that the Sussex had a shorter neck than the other varieties of spaniel. I hasten to point out that this is not so. The old scale of points insisted that the neck should be 'long, strong and slightly arched' and this has never been altered. Montour had a wonderful neck which has, I am glad to say, been handed down. He would probably have been used more if he had not had the misfortune to be undershot. His only other fault was that his coat was not good, being rather harsh in texture. He was an excellent retriever and could be relied upon to pick up anything you asked him to on either land or water. I believe that I have failed to mention that the Sussex has a passion for water, they only have to see a stretch of water and they are into it. This has its drawbacks, especially if you happen to live in a house with a moat!

After the second war Miss Wigg came back into the breed with a brace that she bought from me. She had a lovely dog in Hornshill Duke but he never sired anything and so left no mark on the breed. His litter sister, however, Hornshill Elizabeth went to found Dr. Rickards' kennel. During the time when I was stagnating she had come into the breed and she really did keep the registrations up. In addition, she supplied both the Sedora Kennels and Mrs Elsdens' Chesara Kennels with their foundation bitches.

I used Mrs Elsden's Chesara Pride on my Weiden bitch and from her bred Caribou of Fourclovers who, before he went to America, sired a

45. Tibleash of Fourclovers retrieving his first pheasant.
P. R. Lange

litter which included some useful members which have left their mark on the breed both for work and show.

In Mr Orton's notes recently he refers to some letters written one hundred years ago, which have just been turned up by Mrs Bates. One letter makes reference to the bow-front of the Sussex, but this you never see nowadays. I tried hard to preserve it but the straight-fronted brigade won. They all said that Sussex puppies must not be over-weight, whereas I have always believed that they should be allowed as much food as they can eat. I am sure this craze for straight fronts was the beginning of the downfall of the Sussex. They lost their wide chests and at the same time the beautiful double frill that they once had. They were smooth from the chin down with a short frill either side, rather like a lace cravat. I think the characteristic roll will soon disappear too, as this was obviously due to the wide chest and the tremendous spring of the rib.

The other thing referred to in this letter is some mysterious difference in the tongue. This I have never heard before. I thought that Mr R. C. Morris of Ticehurst knew all the peculiarities of the breed but I never heard him mention any such thing. In the course of fifty years I should have thought that someone would have mentioned it to me. I do wish that the letter had shed some more light on this matter. Probably we shall now never know what was this strange distinction.

I feel that I cannot close this chapter without some reference to the North American Indian, known to us as the Red Indian. I have always had a great admiration for this much abused nation and I have called all my dogs by Red Indian names. In many ways the Sussex spaniel has had similar ups and downs and both have many times been close to extinction and then revived. At present, both seem to be undergoing an upward trend. I hope and pray that this may continue.

11 Field Spaniel

The earliest reference to what we know as the Field Spaniel came from 'Stonehenge'. He was Mr J. Walsh, Editor of *The Field* and author of a book *The Dogs of the British Islands*, 1867. He mentions that among the Norfolk Spaniels and mixed breeds were some black, black and tan and even liver-coloured bred by a Mr Burdett. This Mr Burdett was the Secretary of some of the first Birmingham Dog Shows in the 1860s and it appears that he acquired these dogs from a Mr Footman of Lutterworth whose family had bred them for many years. In this book it is claimed that the same type were seen as early as 1836 and, although rare, they could be procured.

Nevertheless, 'Stonehenge' claimed in his third edition (1878) that much Sussex Spaniel blood was introduced by Mr Phineas Bullock (of Bilston, Staffs) who, with Mr Jones of Oscott, had acquired most of Mr Burdett's dogs after his early death. In this modern age of pure breeding it is not easy to appreciate that the Spaniel breed was divided first into 'land' and 'water' Spaniels, and the former, also known collectively as 'field' Spaniels were divided into varieties—Clumber, Sussex and Field.

The Field of a hundred years ago also included the ancestors of the modern Cocker and Springer Spaniels which, in course of time, became recognised by the Kennel Club as separate breeds. In these early days it was almost common practice to inter-breed among these varieties and to register and show progeny according to the variety most closely resembled. Classes at Dog Shows were provided for some varieties, and it was common for Field Spaniels to be divided by a weight limit of 25 lb. Those under that limit became recognised as Cocker Spaniels, while the liver/white and black/white Springers were divided from these over the limit, and the generally solid coloured dogs left retained the name 'Field Spaniels'.

The breeding with Sussex Spaniels continued into the 1870s with Mr Jacobs of Newton Abbot. He produced it by using a Sussex Spaniel dog 'Champion Batchelor', bred by Dr. Spurgin who was later responsible for establishing parti-coloured Field Spaniels. This Sussex dog, however, was not pure as his breeding included Black Field Spaniels and it was also said that he did not properly resemble a Sussex nor was his colour the true golden liver of that breed.

Mr Jacobs produced his entire strain from four black Fields: Nigger whom he bought from Mr Bullock, and three bitches of Mr Boulton's breeding (Beverley) called Smut, Smutty, and Negress. To improve bone he purchased Champion Batchelor and a Sussex bitch called Russet which he considered had much of the Field type. He produced his first outstanding dog Kaffir (Batchelor × Smutty) and litter brother Zulu in 1878 and because Batchelor was killed by his Gordon Setter Champion Marquis, he proceeded to in-breed. In this he was so successful that his 'Newton Abbot' Field Spaniels were almost invincible in the Show ring until he eventually gave up in 1892.

Mr Jacobs was one of the founders of The Spaniel Club which was indirectly the cause of his giving up his Spaniels. He objected to a rule which stated that where the Club's Special prizes were offered the judge should be chosen from members of that Club. He claimed that 'A lot of mushroom judges sprang up' and 'ringing the changes' commenced in earnest among the exhibitors. He last exhibited at the Birmingham Show 1892 where the judge had never owned a Field and was heard to ask the difference between a Cocker and a Springer!

He decided to give up and sold his last twelve dogs for £1,500. He was certainly the most successful breeder of his time and did most to establish and promote the Black Field Spaniel. He exported many dogs, some of his earliest going to America where they played a large part in setting the type of both Field and Cocker Spaniels.

In the 1870s it had been considered undesirable to breed coloured Field Spaniels and any born were unwanted. However, Dr. Spurgin, who was a successful breeder at the time, persevered, bred, and exhibited these coloured puppies (usually black/white, or liver/white) and is considered as much responsible for the coloured variety as was Mr Jacobs for the solid colours. They were never as popular as blacks, and, worst of all, they developed and exaggerated the weaknesses in the breed that led to

46. Black Field Spaniel in 1880. Mr T. Jacob's Kaffir.

47. A rare picture of a Field Spaniel bitch from the period 1900-1910.

its decline. It was later said to have been proved that they were bred to Basset Hounds to produce better colours and longer backs, and, indeed, the illustrations of the variety at the turn of the century do nothing to disprove the theory. It was said to be vouched for by the Earl of Devon who kept Basset Hounds, and to whose dogs Field Spaniel breeders sent their bitches.

One of Mr Jacobs most noted dogs Newton Abbot Perfection was sold to another breeder Mr Moses Woolland who re-named her Bridford Perfection. The sale price, considered a record at the time, was said to be 270 guineas.

It was from the dogs he bought from Mr Jacobs, that Mr Woolland established his Bridford kennels. It was said that he only exhibited at the biggest shows every year but still took most of the prizes whilst Bridford dogs sold by him were constant winners at smaller exhibitions. His withdrawal from breeding in 1905 was undramatic and caused by pressure of municipal duties in the Borough of Kensington. His whole Kennel was auctioned in July of that year. At a later auction of Black Fields belonging to Mrs F. and Master W. Woolland, the best price was 60 guineas for Bridford Marquis who either then or later passed into the hands of Mr Ernest Trimble. Mr Woolland was the owner of a department store in London and two of his Black Fields which he had sent to a taxidermist, reposed for many years either side of the entrance to the store.

The type of Field Spaniel which prevailed before 1914 deserves consideration. In the early period before Mr Jacobs (i.e. 1870s), it seems that the best Field Spaniels, whilst exhibiting a noticeable length of body in proportion to the leg, were considerably higher on the leg than those that followed them. Whether it was the introduction of more Sussex blood (and later Basset blood into the coloured variety) or a deliberate intention to create a lower and longer dog is difficult to tell. Mr Jacobs claimed that he wanted this new type, but it is not clear whether he bred deliberately for it, or, finding it as a fact, approved of it retrospectively.

Illustrations show that short legs and long bodies were standard and it

48. A Field Spaniel puppy belonging to Mr Mortimer-Smith, photographed in 1912.

is clear from all authorities that this exaggeration was taken to such an extreme as to account for the rapid decline of the breed between 1900 and 1914. Exhibitors at shows were seen to stretch their dogs out as long as possible and judges compared lengths side by side. Crooked front legs and sagging backs were very common although there are illustrations of dogs at this period which, although long, appeared not to have these faults. It coincided with the rise of the Cocker and English Springer Spaniels which, being generally of better conformation, replaced the Field in popularity both as show and working dogs.

Fields were run in trials and were used extensively by shooting men. Besford Bustle and Besford Beauty were successful field triallers owned by Mr Elliot Scott. The need for speed, especially in the Spaniel Trials which commenced in 1899, produced a preference for other breeds. Some shooting men still preferred them for their slowness and their work in thick cover but these were the men who had always had the breed and rejected a change to the faster and flashier dogs of the new century.

Registrations of the breed at the Kennel Club dropped from 286 in 1908 to 125 in 1913. By 1915 the effect of the War had reduced this to 37 and two years later it was only 15. The problems of keeping dogs in wartime contributed so much to the decline of the breed and, coinciding with its general rapid decline in popularity, dealt a savage blow indeed.

C. A. Phillips takes some of the credit for the recovery of the breed because he judged the Royal Show at Manchester in 1916. He gave the Challenge Certificates to Wardleworth Marvel and Modern Maid, both from the same litter exhibited by Mr J. S. Budge and both of an entirely different build from those that were then considered the orthodox type. Mr Budge came from Rochdale and had bred Fields for many years. Wardleworth Sultan had won well for him in the early 1890s. At that time they had a great influence on the breed but it is an influence which continues up to the present day because these Wardleworth dogs produced the foundations of Mr Jones's Darland Kennel, Mr Mortimer-Smith's Wribbenhall Kennel and Mr Kelland's Nobel Kennel to mention but a few. They are, therefore, in the pedigrees of all present Field Spaniels.

This considerable change in the conformation of the breed was almost certainly responsible for the revival of interest in the breed in the 1920s, and in 1923 the Field Spaniel Society was founded.

Mr Mortimer-Smith had bred Fields before the War but started again in the breed after the War when the type had changed. His Wribbenhall Kennel had a great influence on the breed between the Wars. His first successes were with Ch. Wribbenhall Waiter, winner of eight Challenge Certificates in 1925–7 and litter sister Ch. Wribbenhall Waitress who won four Challenge Certificates in 1925–6. The Sire of these two was Matford Dusk from Mr Ernest Trimble's Kennel and the Dam, Wribbenhall Whisper, had as her grand-sire Wardleworth Marcus, sire of Wardleworth Marvel and Modern Maid.

49. A modern Field Spaniel. Mr R. Hall Jones' Sh. Ch. Alice of Elmbury.

Probably his most well-known dog was Ch. Wribbenhall Wetbob, winner of nine Challenge Certificates in 1933–6. This dog seems to have been formed as a centre-point of the pedigrees. His pedigree is a combination of the Matfords, Wardleworths, Fulfords, Darland's, etc. and the frequency of his appearance is startling when tracing pedigrees of present day dogs back some seven generations or so. Mr Mortimer-Smith bred one other dog, Champion Wribbenhall Warlock who won his three certificates in 1931–2. In bitches, his most successful was Ch. Wribbenhall Waterhen, winner of seven Challenge Certificates in 1931–4 and memorable as being the Dam of Ch. Wribbenhall Wetbob. In addition to Wribbenhall Waitress already mentioned there was Ch. Wribbenhall Waterwave, and Ch. Wribbenhall Waterwitch, the former winning five Challenge Certificates just before the last War and the latter winning her three in 1933.

Ch. Wribbenhall Whitewash, was born in 1928 by Deemster of Dela ex Wribbenhall Workgirl. She was the only coloured Field Spaniel to become a Champion after the First War and there is every reason to think that her owner was rightly proud of her. There can be no doubt that the breed owes much to the work of Mr Mortimer-Smith because as the breed declined, both in entries and registration between the Wars, it was he who did more than any to keep it going. He was President of The Field Spaniel Society from 1948 on its revival after the War until his much lamented death in January 1953. His home was at Wribbenhall near

Bewdley in Worcestershire and he was a member of the Borough Council there from 1934 to 1945 being Mayor of Bewdley 1944–5.

Mr Mortimer-Smith had returned to breeding Field Spaniels because he liked the new type. In the 1930s he seems to have altered the type still further for his later dogs seem larger and built more on springer lines than formerly. *Dog World* asked him to write an article on the breed in 1936 and he accordingly approached Mr Ernest Trimble for some photographs of dogs he had bred many years previous so that he could illustrate the difference. Mr Trimble sent the photographs, but refused to have them published if the purpose was to show that the Fields of 1936 were better than the pre-War Fields. He said 'I fear you have gone from one extreme to the other, we certainly had some that were too long and too low but they were not the majority. If you had kept the type of dog you were showing a few years ago, I could have said you were on the right lines'.

Another breeder to exert considerable influence was Mr R. R. Kelland. His Nobel Field Spaniels were consistent winners between the Wars, but principally in the 1920s. He won well with a dog called Black Prince which impressed C. A. Phillips who judged the Kennel Club Show in 1924. He awarded him his first Challenge Certificate and he went on to win three more in succession. His only other dog to obtain a title was Nobel Bang who won five Challenge Certificates in succession in 1936–7. In bitches only one became a champion, Nobel Black Satin who won six Challenge Certificates between 1926–30. As far as pedigrees are concerned the only one to have had any real influence was Nobel Ballistite, sire of Brentor Beauty who in turn was sire of Ch. Wribbenhall Wetbob.

Mr Kelland played an important part in the development and success of The Field Spaniel Society. The Society was formed in 1923 and originally had a very large membership, believed to have been in excess of 70. Mr Kelland was Joint Secretary with Mr William Humphrey in 1924–5, but in 1925 he became sole Hon. Secretary and remained in office until 1948. He was largely responsible for maintaining the Field Spaniel Society's Trials until 1938 when they were last held. He was always anxious to encourage the breeding of self-coloured dogs, and did not like roans because he considered they caused confusion with the English Springer.

Registrations at the Kennel Club had risen as high as 126 during 1925, but by 1931 had dropped to 26 and averaged about 35 per annum until the war. Mr Kelland was greatly concerned at the lack of support for the breed and his comments in the minute book of the Society express it year after year. In 1937 he also became Hon. Treasurer of the Society in succession to Mr Gordon Owen who had held the office since 1926. Mr Owen did much to support the breed at shows during this time, and his most successful exhibit was Woodbell Perfection who won six Challenge Certificates. Indeed at the Kennel Club Show in 1930 he accounted for more than half the entry and won both Certificates with Woodbell Brigand

and Woodbell Perfection. Woodbell Marquis, brother of Woodbell Brigand, was the sire of Wribbenhall Waterhen and thus grand-sire of Wribbenhall Wetbob.

A regular exhibitor and breeder in the five years before the War was Mr A. A. Winfield whose Fieldwin kennel was renowned. He was most successful with two bitches, Fieldwin Topknot and Lich Doss, both winning four Challenge Certificates. His influence on our pedigrees is through Fieldwin Trigger whose dam was Fieldwin Topknot, and sire, Ch. Wribbenhall Wetbob. Trigger was sire of Rothley Rival, Gypsy of Myvod and Keepsake of Pentwyn all of whom are important in the pedigree.

A great deal is owed to Mr and Mrs S. J. Barnett of Breconshire who bred and exhibited both before and after the War. Bobsleigh of Pentwyn won four Challenge Certificates, and a dog bred by the Barnetts, Llanbrynean Barney, owned by Miss Morgan also won four. Llanbrynean Barney was an important influence. His Sire was Ch. Wribbenhall Wetbob, and his Dam, Icy Kiss whose pedigree contained working Springer blood. Barney was bred to Jean of Llandefalls (bred by the Barnetts) also descended variously from Wetbob and Icy Kiss and produced Vandyke of Pentwyn.

Some of these dogs were also experienced working dogs. Nobel Bang, bred by Mr Kelland won four Challenge Certificates and was a regular runner in trials. He won the Open Stake for Field Spaniels at the Field Spaniel Society's Trials in 1934, and was second in 1933 and 1936. He was then owned by Major-General H. H. Maharaja Dhiraj of Patiala, who was very well known at trials for his working Springers. He presented a very large and heavy Silver Challenge Cup to the Field Spaniel Society's Trials in 1928 for Best Field Spaniel.

Mr J. A. Jefferson was another benefactor of a valuable trophy which was for best Field Spaniel bred by owner at the same trials. It is a silver-gilt cigarette box with a model Field Spaniel mounted on the lid and shooting scenes in relief around the sides.

It would be impossible not to mention Major H. Beaumont whose Strouds Field Spaniels practically upheld the Field Spaniel Stakes at the trials in the later 1930s. Particularly well known was his F.T. Ch. Butter of Strouds. His line produced a slightly smaller type of dog than was favoured at the time, although Brush of Strouds owned by Mr Kelland did win a Challenge Certificate at Crufts 1939. The line is preserved in the pedigree through Bourbon of Strouds who, bred to Vandyke of Pentwyn, produced Ballerina of Bableigh. Major Beaumont also presented a Silver Cup to the trials. He won the Field Spaniel Open Stake in 1928 with Nobel Brownie Lass, in 1930 and 1932 with Breckon of Strouds, in 1936 with Brush of Strouds, and 1937 with Brag of Strouds. Other winners at the trials were Mr Jefferson's Roy of Packwood and Tiny of Packwood, and Mr Kelland's Nobel Ballistite already mentioned. Mr Lewis Beaumont's

F.T. Ch. Saxpence of Skerries won in 1938, and represented the breed honourably at other trials during this year. In the Spaniel Club's stake for AV Spaniel except English Springer or Cocker, Fields took the first three places.

It is interesting to note that many of these men who ran Field Spaniels in trials at this time were probably best remembered for their achievements in other varieties; Major Beaumont, Mr Lewis Beaumont, Capt. Tredinnick, Mr Jefferson and The Maharajah of Patiala all kept working English Springers and ran them successfully too, while Mr Kelland kept and ran both English Springers and Cockers. Mr Mortimer-Smith, however, was devoted just to Field Spaniels and ran them successfully in the 1920s both at the Field Spaniel Society's trials and at the Spaniel Club's trials where, between 1926 and 1929, there was also run an open stake for Field Spaniels.

It is impossible to conjecture what future the breed would have had in the 1940s had it not been for the War. It is clear that, although numerically weak, they were well regarded by gundog folk, for in 1937 W. Calvert, writing in *Our Dogs* claimed that of all the spaniel varieties the Field came nearest to the ideal laid down in the standard and he also said 'I find a consensus of expert opinion on the part of unbiased gundog men that we have today Fields which are second to no variety of sporting spaniel in their close approach to the ideal'.

The effect of the War is easily seen in the registration figures. In 1942 and 1943 there were only two dogs registered, and the whole decade did not average 10 per annum. Challenge Certificates were suspended with dog shows in mid 1939 and the last field trial was held in 1938. It was 1948 before the Field Spaniel Society was able to meet again and we owe much to those who gathered then and by careful breeding from the few dogs left, raised registrations as high as 45 in 1950.

With the resumption of dog shows the meeting was held at the West of England Ladies Kennel Society's Show. There were ten members present. Mr Kelland insisted that he could no longer act as Secretary, and Mrs R. Grayson was appointed. Mr Mortimer-Smith was elected President and Mr Barnett Chairman. It was this team that led the revival of the breed in the early 1950s. Mr W. W. Gale succeeding Mrs Grayson in 1952, Mr Penhale succeeding Mr Barnett in the same year, and Miss P. B. Heynes succeeding Mr Mortimer-Smith in 1953. Challenge Certificates returned in 1951 at three shows and, with a break in 1952–3, returned again in 1954–60 although only two shows had them in 1956–7 and 1959–60. Mr J. B. Taylor, however, ran his dog Rothley Bruin of Strouds frequently and with some success, notably at the Spaniel Club's trials between 1948–52, and Mrs Barnett and others also ran dogs in trials.

Entries at the Shows during this period were never high, varying from about 10 to 30 depending on the show. One of the best known dogs was
 Mrs Barnett's Vandyke of Pentwyn who won Best of Breed very often

and the Gundog Group at W.E.L.K.S. Show in 1948. He would surely have obtained a title had there been Challenge Certificates at the time. Mrs Penhale's Brag of Bableigh was a consistent winner and collected eight Challenge Certificates between 1951 and 1956.

Another consistent winner was Mr Jack Tannant's Sh. Ch. Kip of Kiondroughad, winner of seven Challenge Certificates. He was by Renoir of Westwind out of a liver and white bitch Gypsy of Myvod. The sire of this bitch was Mr A. A. Winfield's Fieldwin Trigger who, it will be recalled, was a son of Wribbenhall Wetbob.

Mrs Grayson's Ch. Renoir of Westwind and Miss Haynes Sh. Ch. Teffont Lac each obtained the three certificates for their titles during the 1950s. It is notable that Vandyke of Pentwyn was the sire of Ballerina of Bableigh and Renoir of Westwind, who in turn was the sire of Teffont Lac and Kip of Kiondronghad. Three bitches obtained titles then. Mr G. R. Escott Cox's Sh. Ch. Pirouette of Pentwyn, winner of twelve Challenge Certificates, is the present record holder, in Spring 1974.

Pirouette was by Kip of Kiondroughad ex Ballerina of Bableigh and bred to Teffont Lac she produced both Ronayne Royalist and Columbina of Teffont, both essential elements in the modern pedigree. Other notable bitches were Ballerina of Bableigh also belonging to Mrs Barnett who gathered five Challenge Certificates, Teffont Lalage (Miss Haynes) who won three and Ronayne Georgiana belonging to Mr and Mrs A. J. Everton who just won two before the Certificates were withdrawn in 1960, and unfortunately failed to get her title.

Mr G. R. Escott Cox's Ronayne Field Spaniels did much to secure the future of the breed. He became Secretary and Treasurer of the Society in succession to Mr Gale in 1954 and held that office until 1963 when his wife succeeded him as Hon. Treasurer. His successor as Secretary was Mrs Beryl Everton 'Ridware'. These were the people who carried the breed forward into the 1960s together with Miss Phyllis Haynes's 'Teffont' and Jack Tannant's 'Rhiwlas.' There were others then and previously but the blood lines have died out like Mrs Taylor's Beaucaris Kennel. The last of the Beaucaris Kennel was Beaucaris Beaujason, a handsome liver dog owned by Mrs Peggy Swanson of Guildford, but he died in 1966.

Miss Haynes, who only kept Black Fields, remained the President of the Society until her death in 1962. Of her two dogs, Sh. Ch. Teffont Lac and Columbina of Teffont, Lac stayed in Wiltshire and Columbina fortunately came to Mrs Everton and proved invaluable in the fight to preserve the breed. Miss Haynes was a familiar figure at the four shows that still classified the breed—Crufts, W.E.L.K.S., National and Birmingham.

Mr Escott Cox had bred his last litter some time before, and in that year Mr Tannant bred his last until his recent and very welcome return to the breed. Registrations, which had been increasing in the early 1950s, crept lower and the years 1959 to 1961 produced together only seven registrations. There were none in 1962, nor were there any entries at Crufts Show that year.

In 1963 two litters were registered although they had all been born in 1962. Jack Tannant had a litter of seven from Justice of Rhiwlas ex Jillian of Rhiwlas, litter brother and sister. The author obtained one of these, a bitch, Elmbury Morwena of Rhiwlas, his first Field, and sadly the only one in the litter to survive. The other litter was bred by Mr and Mrs Everton; the dog was Mr Escott Cox's Ronayne Royalist and the bitch an interbred Spaniel, Red Imp of Rhiwlas, and there were eight pups.

It is necessary to explain the 'inter-breeding'. It has been shown that many times Spaniel breeds have been inter-bred, not only in the last century but during the 1920s and 1930s. Icy Kiss, previously mentioned, was produced from a Field bitch Wardleworth Libby, and an English Springer dog, Dismal Desmond. It is most probable that this was done to improve the quality of the breed. Dismal Desmond was of Field Trial breeding by Vim of Invertrossocks ex L'ile Rosebud. These both had interesting sires, Dual Champion Flint of Avondale and Champion Little Brand. Both Mr Kelland and Major Beaumont were inter-breeding and Mr Kelland ran Nobel Neonite (Field × Springer) at Trials in 1929 as a Springer! In the late 1950s it was decided to inter-breed again, but this time to introduce fresh blood into the breed.

The dog chosen for this was Lady Lambe's Whaddon Chase Duke who was bred to Show Champion Teffont Lalage. From the litter, Sherelake of Rhiwlas was bred to Justice of Rhiwlas to produce Red Imp of Rhiwlas, who, when bred back to Field again, Ronayne Royalist, produced the puppies eligible for registration as pure Field Spaniels. In 1962 there was another litter by Ronayne Royalist ex Red Imp of Rhiwlas bred in 1964 but nothing is known of these and they have not been bred from. All modern Field Spaniels are produced from the four dogs mentioned, the two dogs, litter brothers, Ronayne Regal and Gormac Teal, and the two bitches, Columbina of Teffont and Elmbury Morwena of Rhiwlas. The first breeding was from Columbina of Teffont by Ronayne Regal in 1965 and there were six puppies. Mrs Everton kept two, Ridware Empress and a dog, Ridware Emperor, who has twice sired a litter. There was a repeat breeding in 1966 which produced five puppies, Ridware Roana being notable as a blue roan and kept by Mrs Everton, and Sh. Ch. Mittina Ridware Samantha purchased by my mother, Mrs A. M. Jones and bred from twice to date. Also in 1966 Elmbury Morwena of Rhiwlas was bred to Ridware Emperor and produced eight puppies. She had not been bred earlier because it was hoped she would produce a litter by Beaucaris Beaujason. Of the eight puppies three went to the U.S.A. to re-establish the breed there where they had become extinct; they were Brigadier of Mittina who went to Mr P. Carl Tuttle and Pilgrim of Mittina and Flowering May of Mittina who went to Mr Richard Squier. Field Marshal of Mittina went to Mrs Swanson to replace 'Jason' and the remaining four stayed at home, a dog Sh. Ch. Little Corporal of Mittina, and three bitches Vanessa of Mittina, Lise of Mittina and Juno of Elmbury. Mr Escott Cox's Ronayne Royalist was a fine dog and a consistent winner for many years. Columbina, though small, was a sound bitch and did some winning as did Ridware Annabel re-acquired by Mrs Everton and actually owned by David Everton. Beaucaris Beaujason won a Best of Breed at about ten years old; one of the few times he was shown.

1968 brought a useful litter from Mittina Ridware Samantha by Gormac Teal, seven in all and only one dog, Jonathan of Mittina who stayed at home. Jeannie of Mittina and Joanne of Mittina went to Mr Squier and Mr Tuttle respectively in the U.S.A. Sh. Ch. Jillian of Mittina went to Miss Tessa Whitley, whose father Mr J. R. Whitley became Hon. Treasurer of the Society in 1969. Sh. Ch. Juanita of Mittina went to Mr and Mrs Richard Middleton in Yorkshire, and Sh. Ch. Judith of Mittina went to Mr Wilf Foulds in Lancashire. The last one, Jessica of Mittina went to Scotland to a gamekeeper together with Megan of Mittina but when he had to give up about eighteen months later, they both came back home. Megan was from a litter of nine born at the very end of 1968 by Field Marshall of Mittina ex Lise of Mittina, of which Mallard of Mittina belonging to Mr and Mrs Middleton has been used at stud, and Mavis and Milly have been bred from. Mr Middleton has been Chairman of the Society since 1970. The last registrations of the decade were a litter by Gormac Teal ex Juno of Elmbury which produced eight puppies. Three of these stayed at home, a liver dog Ch. Adrian of Elmbury the first for many years to obtain a Qualifying Certificate, also a Junior Warrant and eleven Challenge Certificates. Sh. Ch. Alice of Elmbury, a black bitch, and Little Drummer Boy of Mittina, a blue roan dog. Adam of Elmbury and Anna of Elmbury both went to Mrs E. Roberts of Welsh Springer fame, and Adam was more recently acquired by Mrs Peggy Grayson on her returning to the breed after about 20 years. Of their remaining three dogs, only Anthony of Elmbury is sometimes seen, as his owner is Hon. Auditor of the Society. It was a great occasion when Challenge Certificates were regained in 1969. An alteration to Kennel Club rules provided that twenty-one registrations over a three-year period were sufficient and an allocation of four sets was made for that year. This was increased to five in 1970 in view of the visit of the European Spaniel Congress to the City of Birmingham Show. In 1971 five were allocated again, and increased to six for 1972. By 1972 an aggregate of over seventy-five in three years provided for eight sets in 1973 and this was maintained for 1974.

Three other dogs not previously mentioned have gained their titles; Sh. Ch. Jolyon of Elmbury (Jonathan of Mittina × Vanessa of Mittina), Sh. Ch. Mittina Cleo (Sh. Ch. Jolyon of Elmbury × Lise of Mittina) and Sh. Ch. Dayhouse Barnaby (Templetrine Michael × Anna of Elmbury). Mr L. French has trained his black dog Mittina Nicholas and in October 1972 he won his qualifying Certificate and best performance at the Clumber

Spaniel Society's Qualifying Trials. The Field Spaniel Society has been increasing its membership over the last few years and this now stands at nearly fifty. A Year Book has been produced on a modest scale since 1970 and it was possible to celebrate the Jubilee of the Society in 1973 with a special party and exhibition at W.E.L.K.S. Championship Show.

The future of the breed appears moderately secure provided that interest in the breed is maintained and breeding is continued on the basis of as little in-breeding as possible. The interest in exhibiting the dogs is at present considerable in proportion to the numerical strength, and it is encouraging that some interest in working dogs has been aroused and is being pursued.

The Standard of the Field Spaniel has remained unchanged since revised in the early 1920s when the separate Standards for the 'Black' and the 'Coloured' Varieties were condensed into one and phrases like 'Body long and very low' were removed. It is now intended that it should be altered slightly to include the lighter shades of eye colour to be expected in liver and roan coloured dogs.

The most distinctive characteristic of the Field is the shape and nobility of the head, to quote that standard 'it's very stamp and countenance should at once convey the impression of high breeding, character and nobility'. The head is remarkable in its length and particularly the length of muzzle; it is comparatively fine and has a 'distinct occipital protuberance' formerly described as a 'distinctly elevated occipital tuberosity'. This 'protuberance' is curiously similar to that found in some Hound breeds like the Basset; but they should be similar in no other way, a coarseness or flatness of the skull is not desirable. The muzzle should be long and lean, neither snipy nor squarely cut, for the purpose as set out in the standard is to develop the olfactory nerve and attain the best scenting powers. The ears will not be unlike those of a Cocker in shape but there is a tendency to find some of very considerable length which, with a long muzzle, does tend to achieve a greater balance and pleasing aspect.

Eyes should be moderate in size, and show no haw. Colour is of some importance and they should basically match and correspond with the colour of the dog. It is rare that they do not in fact do this. The most usual colours are dark brown or nearly black in black coated dogs, and hazel-brown in liver coloured dogs. The teeth should be level and strong, and the neck long, strong and muscular both for the purpose of enabling the dog to retrieve with ease and carry game well.

It has been seen above that the proportions of the dog particularly concerning height at shoulder and length of body have always been a matter of controversy. The answer must lie in moderation. The 1920 Standard said 'body of moderate length'. Mr Trimble condemned the 'caterpillars' but also the semi-English Springers. The best type will lie between the two. The Field can compete with the English Springer for speed and efficiency but each Variety of Spaniel has properly a different role to play, and is built and bred to be suited to that role. Some are built for speed in open country, others for hunting in dense cover. The Field fits between these extremes, he is built for moderate speed, but he is heavily built to assist in the penetration of cover. Thus the proportions between legs and body should be comparatively short and comparatively long, but never so much as to produce any weakness in the back. A deep chest is typical of good specimens of the breed.

The Standard provides for a flat or slightly waved coat, never curled like a Water Spaniel. Texture is usually silky as provided for, and feathering whilst always abundant in certain parts, is often sparse under the chest. A dog which is working in bracken will quickly lose almost all feather, but this applies to any breed and is therefore a matter of little merit.

The tail is always carried low or sometimes level with the back. It is always very merry in action and although not provided in the standard it should always be docked. Normally docking is taken to an unnecessary extreme in that it is usual in this breed and others like Cockers to dock off two-thirds of the tail. In this breed it has a much more pleasing effect to dock not more than half and only one-third is really necessary. A dog with a length of body is better balanced with a longer tail and only one-third is necessary to prevent laceration in cover.

Colour has already been mentioned. Any colour is allowed but certain colours, Black and White, Liver and White, Red or Orange and White, are supposed to be considered faults. The purpose of this was almost certainly to keep some difference between Field and English Springers when the two became rather more similar in type. It is doubtful if there is any merit in it now. The usual colours are Black or Liver, and these colours will account equally for about 95 per cent of all puppies born. The incidence of other colours has been higher in U.S.A. The other colours known in the U.K. are liver roan and blue roan. Some Blacks and Livers have some traces of tan markings but the only properly marked black and tans are in U.S.A. A certain amount of what the Americans call 'Irish Spotting' is also usual in the breed, indeed there are few solid coloured dogs that do not have some white markings on the chest, etc. The usual appearance is a few white marks on the front of the chest and the greater the influence of this gene the more white, i.e. it spreads upwards over the under part of the neck, the chin, and finally in most extreme cases onto the top of the muzzle and also appears on the feet. This might be considered undesirable but a white nose is an advantage for a working dog and it hardly seems a matter for concern even in extreme cases.

The Field Spaniel makes a willing worker and is easy to train especially in the early stages. He has a certain independence that needs to be controlled, but his willingness to please and his apparent delight in his work is certain to satisfy his owner. He also makes a delightful companion and the ideal pet for a country home. Experience has shown that children and this breed get along very well together and are soon devoted to each other.

They are renowned for their friendliness and reliability and some have even succeeded in competitive obedience work.

The breed has nearly been 'down and out' several times. It is now enjoying a new lease of life which it is vital must be retained and the progress made consolidated to ensure the future of the breed. Once you have met a Field Spaniel you know it is worth the effort, whatever the cost, to be sure this breed is safe and never threatened with extinction again.

12 American Cocker Spaniel

The American Cocker Spaniel and its cousin the English Cocker were evolved from the same basic stock. Up to the 1870s the Spaniel family were very mixed in England and small spaniels were known as Cockers regardless of how their parents were described. Cockers and Fields came from the same litters, and the only qualification a dog required to compete as a Cocker was that he should be under the weight of 25 lb., a limit somewhat irrationally fixed since in the case of an animal just on the border line he was a Cocker in the morning and a Field after a good dinner.

In the year 1879 the most famous of all Cockers was born. His name was Obo and he was the offspring of a Sussex sire and a Field dam, the breeder being Mr James Farrow. His son, Obo II was born in Canada out of a bitch called Chloe II who had been exported in whelp by Mr F. Pitcher, and he was sold to Mr J. P. Willey of Salmon Falls, New Hampshire.

Obo II weighed $23\frac{1}{2}$ lb. and measured $9\frac{1}{2}$ in. from foot to withers; he was black and carried a curly coat especially on the shoulders and hindquarters, and had profuse feathering. He was the prince of stud Cockers and his true value was the blood lines he passed on to posterity. He was the direct ancestor of Robinhurst Foreglow, and all winning Cockers of yesterday and today are direct descendants of four sons of Foreglow. They were Red Brucie, Ch. Sandsprings Surmise, Ch. Midkiff Miracle Man and the Canadian Ch. Limestone Laddie. All other lines not having this blood have completely disappeared from the breed over the years. Robinhurst Foreglow was owned by Judge Townsend Scudder. He was a 28–30 lb. dog of good substance. Blackstone Chief the sire of Foreglow weighed 26 lb. and was lighter in bone.

The story goes that Judge Scudder wanted to buy him, and in order to do this he had to buy the entire Blackstone Kennel—32 dogs in all! He offered Foreglow at stud without a fee to anyone owning a typical Cocker bitch and desiring to breed to the dog. Many of the breeders of that time, not being accustomed to the type and size of Foreglow hesitated to take advantage of this offer, and continued to breed with their own dogs who were long in the back and short on the leg. Two breeders, more astute than the rest, took advantage of this offer realising that there was about to be a change in the appearance of the Cocker Spaniel of that time.

They were W. T. Payne of the Midkiff Kennels, Pennsylvania, and Herman Mellenthin of Poughkeepsie, N.Y. and both these gentlemen wrote history in the book of the American Cocker. Mr Payne of the Midkiff Kennels was the first man to sell a dog at a price running into four figures—this was the price he got for Ch. Midkiff Miracle Man whose dam was a Best in Show winner at the Westminster Show, New York in 1921.

In the 1930s the American Cocker really began to go places, and registrations shot up at the American Kennel Club. Some of the great stud dogs of the period were Ch. Stockdale Town Talk, a black whelped in 1939, Ch. Maddies Vagabond Return, a buff whelped in 1949, Ch. Honey Creek Heirloom, a parti-colour also whelped in 1949, and Ch. St. Andrea's Medicine Man, black and tan, whelped in 1950.

In those early dog show days the only trimming allowed was the plucking by thumb and finger on the animal's head. The first really trimmed Cocker was exhibited by W. T. Payne at Westminster Kennel Club Show at the old Garden in New York. Here originated the vogue of clipped heads and necks that created such a sensation amongst the breeders. At that time American Cockers were shown in the English fashion and posed naturally. The practice of highly specialised posing and gaiting came much later.

In the late 1890s and 1900s American breeders were rigid in breeding small dogs weighing not more than 24 or 25 lb., looking upon it more as a house pet, but those who wanted a sporting type of dog required a heavier animal and these contentions caused considerable friction at the Shows. As a result of this weight problem an exhibitor had the right to challenge the weight limit of an opposing dog, and when the standard reduced the weight limits, scales were introduced into the ring. If a dog was over the weight limit he was disqualified and not permitted to compete.

As the popularity of the Cocker increased, the winning dogs from Canada gave way to those bred in the United States. In 1921 a Cocker Spaniel was the winner of Best in Show at the Westminster Show at the old Madison Square Garden in New York. This was W. T. Payne's Midkiff Seductive, a black and white, and not until 1940 and 1941 did another Cocker hold the spotlight at another Westminster Dog Show, this time at the new Madison Square Garden.

The breed continued to change and develop, and public admiration for a type, its acceptance by leading judges—its exhibition by top breeders—all of these factors served to change the accepted type and produce the modern American Cocker. The real break in Cocker type came in 1920; they were taller, more upstanding at the shoulder, coats were longer and
 feathering more profuse, and stifles more angulated. They were no longer

acceptable undersized and low to ground.

The breed in America had begun as a continuation of its English counterpart, and as less were imported a type was bred entirely by selection to fit American ideals and not by the use of foreign blood lines. American and English Cockers diverged so widely that they ceased to be recognisable as the same breed. In 1936 the American Kennel Club recognised this divergence and established classes for what were called English Type Cocker Spaniels as a separate variety.

In 1945 the American Kennel Club declared the English Cocker Spaniel as a separate variety and gave it a breed status, as distinct from the breed known in America as the Cocker Spaniel.

The correct trimming of the coat of the American Cocker has become an art and clipping and thinning is now essential. It is necessary for the exhibitor to remove long hair from the parts on which it is not wanted with the use of clippers, and even the pet owner needs to have this done frequently by an expert. The American Cocker should be merry with no suggestion of timidity. Resentment to handling or bad treatment should not be construed as bad temperament nor should reaction to fright as timidity. One of the greatest charms of the breed is the gentle expression of the eyes so characteristic of the breed.

The general appearance of the modern American Cocker is that of a serviceable looking dog, with a refined, chiselled head, standing on straight legs and well up at the shoulder, compact body and wide muscular quarters. The dog's sturdy body, powerful quarters and strong well-boned legs show him to be capable of considerable speed combined with great endurance. Above all he must be free and merry, sound, well balanced throughout, and in action show a keen inclination to work, be equable in temperament and with no suggestion of timidity. The head and skull should be well developed and rounded with no tendency towards flatness or pronounced roundness of the crown. The forehead should be smooth and free from wrinkles, the eyebrows and stop clearly defined. The bony structure surrounding the socket of the eye should be well chiselled; with no suggestion of fullness under the eyes nor prominence in the cheeks which, like the muzzle, should present a smooth, cleancut appearance. The muzzle should be broad and deep with square even jaws. The upper lip should be of sufficient depth to cover the lower jaw, presenting a square appearance. The nose must be of sufficient size to balance the muzzle and foreface, with well developed nostrils and black in colour in the blacks and black and tans; in the buffs, parti-colours and roans it may be black or brown, the darker colour being preferable. The teeth should be sound and regular and set at right angles to their respective jaws. The relation of the upper teeth to the lower should be that of scissors with the inner surface of the upper in contact with the lower surface of the lower jaws when closed. The eyeballs should be round and full and set in the surrounding tissue to look directly forward and give the eye a slightly almond-shaped appearance. The expression should be intelligent, alert, soft and appealing.

The ears should be long, set on a line no higher than the lower part of the eye, the feathers fine and extending to the nostrils; well clothed with long, silky, straight or wavy hair. The neck should be sufficiently long to allow the nose to reach the ground easily. It should be muscular and free from throatiness, rising strongly from the shoulders, and arch slightly as it tapers to join the head. The shoulders deep, cleancut and sloping without protusion and set so that the upper point of the withers is at an angle which permits a wide spring of rib. Strongly boned, straight and muscular forelegs, set close to the body well under the scapulae, the elbows turning neither in nor out and the pasterns short and strong.

The height of the dog at the withers should approximate the length from the withers to the set-on of the tail. The chest deep, its lowest point no higher than the elbows, its front sufficiently wide for adequate heart and lung space, yet not so wide as to interfere with straight forward movement of the forelegs. Deep ribs and well sprung throughout. Short body in the couplings and flank with the depth at the flank somewhat less than at the rib. Strong back, sloping evenly and slightly downward from the withers to the set-on of the tail. Wide hips and quarters, well rounded and muscular. The body should appear short, compact and firmly knit together, giving the impression of strength. The hind legs should be well boned and muscled with good angulation at the stifle and powerful, clearly defined thighs. The stifle joint should be strong and there should be no slippage in motion or when standing. Hocks strong, well let down and when viewed from behind, the hind legs should be parallel when in motion and at rest. The feet compact, round and firm with deep strong tough pads and hair beteeen the toes. They should turn neither in nor out.

The docked tail should be set on and carried on a line with the top line of the back or slightly higher, never straight up like a terrier and never so low as to indicate timidity. When the dog is in motion the action should be merry. The coat on the head should be short and fine, on the body of medium length with enough undercoating to give protection. The ears, chest, abdomen and legs should be well feathered, but not so excessively as to hide the American Cocker Spaniel's true lines and movement or affect his appearance and function as a sporting dog. The texture is most important. The coat should be silky, flat or slightly wavy, and of a texture which permits easy care. The coat should be neither excessive, curly nor cottony in texture.

The American Cocker colours are black, any solid colour other than black and this includes black and tans, and these are termed ASCOB, and lastly parti-colours. The black should be jet black with no shading of brown or liver in the coat. Black and tan should have definite tan markings on a jet black body. The tan markings should be distinct and plainly visible, the colour of the tan may be from the lightest cream to the darkest red colour. The tan markings located as follows:—1. A clear spot over each eye, 2. On the sides of the muzzle and on the cheeks, 3. On the underside of the ears and tail, 4. On all feet and legs, 5. On the chest. The parti-colours are red/white, black/white, tri-colour and roans of any of these colours.

The ideal height for an adult dog is 15 inches at the withers and for an adult bitch 14 inches. The height may vary one-half inch above or below this ideal. The height is determined by a line perpendicular to the ground from the top of the shoulder blades, the dog standing naturally with the forelegs and the lower hind legs parallel to the line of the measurement.

The American Cocker possesses a typical sporting dog gait. A prerequisite to good movement is balance between the fore and hind quarters. He drives with his powerful rear quarters and is properly constructed to the shoulders and forelegs so that he can reach forward without constriction in a full stride to counter-balance the driving force of the rear. His gait is co-ordinated, smooth and effortless. The dog must cover the ground with his action and excessive animation should never be mistaken for proper gait.

When trained to the gun, the American Cocker soon proves his ability to work. History was made in 1972–73 when Mr R. H. Wylde of the Eldwythe Kennels, Shropshire, trained and qualified in the field two American Cockers at the age of three and five years respectively, and both have since attained the title of full Champions. His older bitch, Ch. Eldwythe Timothy's Black Tiffany was imported in whelp from Canada and was from completely show stock; she was not put to training until turned five years of age but within months ran in the qualifier with considerable success. Her daughter Ch. Eldwythe Enchanto was born in quarantine and commenced training at the age of three; she soon qualified and became a full Champion. To date these are the only American Cockers to have achieved such honours. The American Cocker makes an ideal family pet, he is affectionate and easy to house-train. He soon accommodates himself to all sorts of human and canine environments, he is alert, attentive, loving and loyal and always eager to please. He will take to obedience training like a duck takes to water and he learns rapidly and retains his education.

The breed was first introduced into England from Holland in 1965, but since then many have been imported from the United States. The most important of these being Am. Ch. Dreamridge Dinner Date, winner of the Gundog Group at Crufts in 1972. Other prominent imports have been Am. Ch. Lochranza Evdon's Escort and Am. Ch. Cholsey's Orients Secret Pleasure.

Owners of American Cockers as pets should appreciate that with the heavy coats that most of the breed carry, especially the blacks, daily grooming is necessary and a professional trim will be required at least every six weeks. A steel comb and a wire pin brush can be purchased from

50. American Cocker: Mittina Love me a Little.

most good pet shops and using these daily will help to keep the dog looking his best; in addition a healthy dog whose coat is cared for will never have a doggie odour. The breeder from whom a puppy is purchased should be able to advise on proper coat care and will either trim the dog professionally or supply the name and address of the person nearest to the owner's home address who will be able to undertake the service. When buying a puppy it is wise to see the parents and the litter, handle them and watch the puppy for any tendency to shyness. Many bad dispositioned dogs are the result of bad owners, but others are the product of indiscriminate breeding and rearing.

13 English Setter

Working English Setters inspire the greatest affection and enthusiasm. Their silky coats and feathering, their affectionate disposition and their passion for hunting endear them to their owners, while their stealthy gliding action over the ground make them a pleasure to watch on the blankest day.

HISTORY

The history of English Setters is difficult to determine. Setting dogs were used to locate game from very early times. These first setting dogs were called spaniels. In 1016 King Canute prohibited the keeping of spaniels and greyhounds within a forest. The book, *The Master of Game*, written by Edward Duke of York who was killed at Agincourt in 1415, describes spaniels as being used to flush game for hawking and also for pointing partridges and quail so that a net could be drawn over them. The Duke also stated they were called spaniels as they came from Spain.

A letter dated 1696 in the county archives in Shrewsbury from John Woodhouse to Sir John Leverson-Gower mentions that he had found a setting dog for Sir John, having previously refused one for £8. As agricultural wages were about 10s. per week in 1765 and they would have been less at this time, the price would have been about the equivalent of six months wages—£400 or more today. Dogs specifically used for setting are mentioned in a statute of Queen Anne (1702–1714).

It would seem that by 1700 spaniels with special setting ability were well known and valued. Edward Laverack started breeding them in 1822 and General Hutchinson, in his book *Dog Breaking* (1848), describes them in detail. Edward Laverack in his book, *The Setter*, written in 1872, describes nine strains of setters. The strains are interesting and some now quite lost:

1. The Naworth Castle: Liver and white, broad chested, the property of the Earl of Carlisle. Their distinguishing characteristic was a tuft of hair on the crest. This can still be seen in some working setters.
2. The Featherstone Castle: Liver and white and small.
3. Lord Lovat's breed: Black and tan.

51. Mr R. L. Purcel Llewellin at a Field Trial in 1915.

52. A handful of Setters. Mr W. Humphrey at the English Setter Club Field Trials in 1937.

4. The Earl of Seafield's breed: Black, white and tan and lemon or orange and white.
5. Lord Ossulston's: Jet black.
6. Mr Lort's of Kings Norton, Nr. Birmingham: Black and white, lemon and white.
7. The Llanidloes (Welsh) breed: Milk white or chalk white and rather small.
8. The Russian Setter: The Prince Consort was very fond of this long coated and wilful breed.
9. The Gordon Setter: Black and tan, but in Mr Laverack's time the Duke of Gordon only used black, white and tan, produced by using black and tan dogs on white and black bitches. The colour was easier to see on the hill. At a sale in July, 1836, eleven of these setters averaged 36 guineas. It is interesting to note that the agricultural wage was 10s. per week or about £36 per year. In terms of today's money this means an average £1,000 each.
10. The Irish Setter: Red in colour but of two shades with some white.
11. Mr E. Laverack's Setters: These dogs were ideally blue or red Beltons, that is dogs with a white background and densely ticked with black or lemon. These are the foundation of the show strain.

The Laverack strain passed to Mr R. L. L. Purcell Llewellin and were separately registered as Llewellin Setters in the U.S.A. In the States they had a great vogue from 1896 to 1908. (See A. F. Hashwalt's writings.)

English Setters and all the locating breeds had their hey-day about the 1850s. Everything was in their favour, guns were mainly muzzle-loading with a slow rate of fire, for breech-loaders had yet to achieve general acclaim. Turnips and mangels were the great winter food for stock and ideal for dog work, while ploughing with horse teams left the stubbles far into winter. Only the moors have remained unchanged and are still ideal for dogging.

Everything in those distant days combined to make dogging the best and most pleasant method of shooting. The talk was of how the old dog held his point rather than how many birds were shot. But with the breech-loader, changes in farming and scientific rearing, driving became the *modus operandi* and the use of setting dogs declined. Now the wheel has turned full circle for rising costs are making them attractive once more in suitable places.

Between January, 1947 and January, 1972 in 195 trials held in England, excluding Ireland, Scotland and Wales, the winners of the all aged open stakes were 88 Pointers, 87 English Setters, 21 Irish Setters, 5 Gordon Setters and in four trials no first prize was given. A random selection of more than 200 dogs entered suggests that Pointers were 40 per cent of the entries, English Setters 30 per cent, Irish Setters 20 per cent and Gordon Setters 10 per cent.

There were six professional handlers employed during this time, who worked for a large kennel keeping a single breed, four for pointers and one each for English and Irish Setters. These men were, or are, masters of their craft and enjoyed advantages seldom available to the amateur handler-owners, so it would appear that the ability to win trials would seem to be very even between the four breeds of pointing dogs.

The best English Setters display great bird-finding brains, a combination of an exceptional sense of smell and the intelligence to gauge the position and the distance of their quarry. Good scenting conditions and bad come alike to such dogs and they always display the ability to bring the guns to a good position for the flush and shot.

USE

Recent changes in farming and increasing expense has made driving less popular in places not ideally suited to game birds. If birds are few in number driving can become a cold and boring occupation. Given suitable open country which allows the Setter to be seen (kale and woodland are unsuitable), great fun can be had with these dogs.

The points allow for selective shooting, large coveys which contain young birds suitable for breeding stock can be left unharmed and small coveys, often collections of old birds, can be culled. Hen pheasants can be allowed to go unmolested, while snipe and woodcock can be found in suitable places.

Unfortunately a high standard of dog training is needed for this shooting. A retriever can be anchored to the owner and only let loose when required, but the locating dog must be loose and far away to be of any value.

It is pointless to use these dogs to attempt to find game for a line of walking guns; their real purpose is to bring one or two guns into position to take the shots which would otherwise be shared by a line of six or seven walking guns. Let us consider just how a setter works.

Seven guns walking 45 yards apart cover a front of $45 \times 7 = 315$ yards, with six spaces between the guns and two half spaces at the end of the line, whilst a setter can range out 150 yards on either side of its handler, so covering the 300 yard front and bringing the handler into position to shoot anything in that area. If the gun walks two miles an hour and the dog does 20 miles an hour, the dog goes ten times as fast, so while the man walks 25 yards the dog can go 250 yards, i.e. 112 yards out, 25 yards forward and 112 yards back and then repeat the cast on the other side.

This would give a very close pattern of hunting, which is usually exceeded. If the distance between the casts is 50 yards, the dog goes out 225 yards, forward 50 yards and back 225 yards giving a front of 450 yards. This would be rather too great an advance and might lead to birds being missed. These beats are illustrated in the diagram, see page 199.

So far an attempt has been made to describe the development and use of the English Setter and its attractions and abilities, now we must examine the management of this dog.

CARE

A kennel and run are essential; the kennel must be warm, dry and comfortable with a bed raised 2 in. from the floor. The bed should have sides at least 6 in. high and have a floor and sides with a double thickness of wood to give insulation.

Sawdust, wood shavings, straw or a blanket are needed in the bed and sawdust or wood shavings on the floor of the kennel. The kennel should be draught-proof and at least 9 ft. square. A stable or out-house is excellent and simply needs a suitable bed.

A run of at least 18 ft. in length and 6 ft. wide is needed. The fence round it should be dog proof, 7 ft. high and with an overhang on the inside to prevent climbing out. It is unbelievable, until actually seen, how an active dog can run up wire netting and escape. Dogs can get some exercise in the pen, whilst puppies can run themselves to skin and bone in a quarter-acre enclosure.

Having made certain that you have somewhere to keep an exceedingly active and energetic dog, an English Setter can be brought home and trained.

TRAINING

The first two weeks, however old the dog is, must be given up to making friends with him. He needs to be brushed, combed, patted, talked to and, if young enough, carried about.

Get him friendly and accustomed to your presence and contact. During

53. Slipping the 6th brace at the Kennel Club Field Trials at Orwell Park in 1949. (Left to right) Mr T. Gaunt (Judge), Mr W. Evans, Mr W. East (judge), Mr G. Conway-Williams (referee) and the author.
Sport and General

this time he should be exercised on the lead or in an enclosed space from which he cannot escape. Puppies have to be taught to lead and not struggle against the restraint. Be gentle, patient and quiet and about half an hour will get him running and pulling happily on the lead. This period of acclimatisation is essential for all strange dogs, however old.

Puppies should be fed more often than adult dogs. The routine is: up to 12 weeks, four feeds per day, from 12 to 20 weeks, three feeds, after 26 weeks, once per day. Water should always be available during the day, but is apt to be upset at night and is best removed.

Exercise varies with age but an honest hour's walk will keep a grown dog fit and well, even on the lead.

Having got the dog friendly and walking on the lead, training can be started if he is over 16 weeks old. During these first two weeks a certain basic training has occured naturally. The dog knows his name and has been checked by the word 'No' when doing something wrong. He should come when called—at least when in the mood!

At this stage a Glossary is needed:

1.	Beat	The area and/or the method of hunting.
2.	Flushing	Making birds fly either at command or by stopping too late.
3.	The Gun	The man hoping to shoot, not necessarily the handler.
4.	Handler and handling	In the literal sense the person who pats, combs, feeds and leads, and also in one sense directs the work of the dog in the field.
5.	The Haunt	Place from which the birds flew.
6.	Larking	Hunting larks or small birds and therefore not attending to business.

7.	Nose	Ability to find birds, not entirely sense of smell but partly 'bird-finding brains'.
8.	Stickiness	Not moving freely into flushed birds after a point.

In order to understand the training method used and indeed the use of setters it is essential to grasp what is the object to be attained. Setters are locating dogs used to find game by covering the ground with a regular and consistent pattern of hunting across the wind. They should advance, hunting the ground to the right and the left at such a speed that the gun can walk steadily and quietly forward.

The faster the dog, the wider his beat, allowing him to cross in front of the gun on each half cast. (See diagram on page 199.)

When the dog locates game he indicates its presence the instant he is aware of it and remains stationary on point until asked to advance or he drops automatically to wing. Pointing should over-ride dropping to wing and shot, if the dog knows more game is present. Many of the best dogs do this or stand up slowly after dropping, a most hopeful indication of more chances to come.

Don't ask him to retrieve. He deserves a rest after the shots. All these manoeuvres should be conducted in silence. Even the guns should not talk.

At about four months (although two or three years is not too late), the following lessons should be started. They must be indoors and for this the dog must be accustomed to a room, shed, or garage. A comfortable warm room helps the handler greatly. Messing about with a cold wet

54. Dr J. B. Maurice handling five Setters at Euston, Suffolk in 1951. *C. M. Cooke*

string in a damp barn in January is not conducive to regular patient training. It's most helpful if lessons are regular, and essential that they are short. Ten minutes at a time is enough. The dog's concentration fails after a very short period.

The first lesson is to get the puppy to stop when eager to rush forward. In fact *as* he rushes forward.

Tie a string about 12 ft. long to his collar or neck and get something he likes to eat. Cheese is simple to cut and is clean, for often dog-biscuit is not sufficiently attractive. Get him used to trailing the string, and then show him the bait saying 'Dead, Dead' and throw it underhand along the floor.

These lessons must be conducted alone with the pupil. He will soon rush forward and eat the bait. When he does this enthusiastically—two or three throws should be enough—stop him a yard short of the bait by holding the cord and say 'Toho', count 30 in your head, say 'on on' and let him eat it. After surprisingly few lessons he will learn to stop on the word. Remember he must start after the bait and then stop, not sit beside you waiting for the command 'on on'.

The whole point of the lesson is to stop him in full career—in other words, when he has smelt game and not before he has started to hunt. He should enjoy this lesson: Don't be too rough.

The check cord can be used gently, either by slowly bringing him to a stop, or by jerking him on his back and frightening him. It is the severest weapon in your armoury. I think the handler must believe the dog wants to please and will do so once he has grasped what is wanted. I believe that my failures in training are due to over-severity.

By throwing the bait right and left as well as forward underhand, you will teach him the signs for right, left and forward. In a month he should follow the hand signal, stop without the cord and go on at command.

The next lesson is the 'drop'. Press him down in a sitting position with your left hand, raise your right hand, and say 'drop, drop'. He will gradually learn to go down on the command and finally the hand-signal alone. It is worthwhile to get him to lie down after he sits well.

Gradually get him to drop farther and farther away from you. In time he should drop at 60 or 70 yards.

If he does wrong and then drops never hit him or punish him (I think shaking a dog is worse than hitting him), or you will lose the drop. Once he is dropped, there is nothing to say or do. It requires some self-command to remember this.

When he will drop at some little distance, clap your hands above your head, hold up your right hand and say 'drop'. Suppose that you will shoot from the right shoulder then after the shot you can most easily hold up the gun with the right hand. He will learn to drop at the noise of the clap. All this is best started and almost perfected indoors with the check-cord held or trailing.

55. An English Setter pointing. F.T. Ch. Downsmans Bracken.
C. M. Cooke

As he gets stronger, let him run about out of doors trailing the check-cord and go through the lessons in a field, whilst a shot from a starting pistol can be used instead of clapping your hands.

The lessons can go on concurrently, but make sure the 'Toho' exercise is perfect.

A plastic cord which does not collect the wet is best for out-door work. Use thick leather gloves because, if he pulls at you, the cord will burn like fire.

At about six months old he should be good at all the above lessons.

Now let us summarise his 'vocabulary':

1. Name.
2. No.
3. Dead Dead—Now to work.
4. Toho—stop.
5. Left—hand sign.
6. Right—hand sign.
7. Forward—hand sign.
8. Drop—word, sign and noise.
9. With 4 and 8 a silent dog whistle can be used so no voice is needed.
10. Ordinary whistle on a recall signal instead of 1.
11. On. Replaced by a special whistle signal.

Whistle signals: For recall have a repeated warbling note. For 3 and 8 repeated short blasts of the silent whistle. A single 'Tweet' 'Tweet' with the appropriate hand-signals, for right, left or forward direction signals.

The whistle signals always need an appropriate hand-signal. You will see that the hand-signal for drop and the whistle can be used together. Various recall signs are used. Kneeling and opening the arms and calling his name usually brings a puppy crashing against your chest. A whistle

10. Mrs Freer with two of her Sussex Spaniels.

11. Field Spaniel.

12. Mr T. Spark with two of Capt. W. Parlour's Setters at the Kennel Club Spring Trials, Newmarket, 1973.
Jim Meads

13. Toberdoney Chaucer, who won four stakes out of five appearances as a puppy.

14. Gordon Setter.
G. L. Carlisle

56. The authors' F. T. Ch. Downsman Tormentil, winner of eleven competitions, shown at the Southern and Western Field Trials, Sutton Scotney in April 1974.
Peter Moxon.

and the gesture will do the same. Beckoning, waving your hand down to your knees and calling will do as well. The hand-signal and hand-clap for the drop can be replaced by a shot. Be careful not to fire too close to him and keep the gun or hand raised.

Several things must be noted: All commands are signal words, none are likely to be used except by the handler. 'Drop' is better than 'Down' for this reason. People say 'Down' if a dog jumps up. 'Toho', the key word, is also a nonsense word and never used out of context.

It is hoped that he will have learnt this vocabulary before he starts hunting too seriously, for in this way he learns what's wanted before being tempted too much.

He should now come when called, go right or left or forward, halt at command, lie down at command. You should be able to get him to pass in front of you by first calling and then giving the appropriate direction signal. From this stage some dogs pass straight into the top rank of shooting dogs at about 15 months; they watch the handler, listen to what he says and use their heads. Some, however, become over-excited and lose all contact with the handler. I believe that perfect grounding in these first lessons is the answer.

While the indoor training is being done, your dog must be exercised, taken for walks on the lead and allowed to romp about loose in the open in suitable game-free areas.

As soon as he starts chasing and finding small birds, try to get him to range across in front of you and check him from chasing them. A long check cord may be necessary, but remember a dog that's never loose never comes back voluntarily! If he points, check him with 'Toho', let him road up and flush and drop even if it's only a lark.

Now you must wait patiently until weather and circumstances are favourable. Try and get some ground with birds and choose a day with a warm, moist wind. He is almost sure to point.

If he does not, wait for a month and try again. If the 'Toho' is well drilled into him he can be stopped when his manner shows he has smelt something. Only harm comes from working puppies on barren ground or very sparsely populated ground. Three or four points in an hour and your dog is made—a flush once a fortnight and he may be ruined. Have the check cord trailing, let him range and find. Hold him on point while you count one hundred, then say 'on' and let him road and flush and drop. Try to press him down on the word 'drop' and then again count one hundred before the next move which is to advance to the haunt and allow the dog to sniff about. A couple of points from an already indoor-trained dog and he's 'made'. It is quite astonishing how well they behave and grasp the object of the exercise. After birds have flushed take him to the haunt. It seems to satisfy him.

You may find things more difficult in the dog's second season but if you encounter trouble always return to the first lessons and then work up again. Almost all faults are due to bad management at this stage. You must be certain of your indoor training and your drop at a distance, and you must also get the check-cord drill going well. He can trail the cord now he is full grown; all the time if necessary.

Always work him dead into wind, don't let him go too far left or right and, above all, don't let him get too far in front, boring into the wind. It's easier to control a dog at twenty yards than at a hundred yards.

Three things are fatal—too wide a range, a tired dog and public demonstrations.

In the final training some things are very important. First trust your dog if he points and keep on working. He can smell the game and you can't. Never give up if he thinks he can find.

Next, help him with his beat. His horizon is limited by his height and sometimes you can steer him into birds: A couple of successes and he will watch for help and this may be invaluable.

'Stickiness'—failure to go up to make birds fly is most difficult to overcome. Constant gentle leading or pushing on may overcome this, but sticky dogs often become unsteady to wing.

Every dog must be studied individually and treated accordingly. Some are very late in pointing but become excellent workers in the end.

It is possible to get dogs to point 'dead'; that is to point the game killed so allowing the gun to pick it up. Towered birds (birds which fly vertically after being shot and fall dead at some distance) can be found in this way. Always watch out for a dog being unusually sticky after shots have been fired for you may see a dead bird just under his nose. Don't hunt runners if he points one, far better to shoot it on the ground.

If you have a retriever out insist on it being on the lead for a bit. It may run into shot or, worse still, to point, and usually has to be constantly nagged at to be kept from wandering about.

All these things may lead to disaster and spoil a good setter. The retriever owners are often deeply insulted, but an explanation of the danger of unsteadiness to the retriever, due to the setter hunting, may help you.

This is, to many people, the most fascinating form of shooting, requiring woodcraft on the part of all the guns, though admittedly the average shot is not too difficult.

Unfortunately, especially on the moors, guns tend to walk the contours of the ground regardless of the wind. If grouse are plentiful, enough shots are obtained to keep the guns contented but the dogs are usually ruined. The term a 'cheek wind' is used to mask this lazy manoeuvre. A dog can work wind of 45 degrees to the handler but not a true sidewind, one of 90 degrees. (See diagram on page 199.) Try to give the dog every chance to excel however hard the walking.

A dog can work down-wind, looping out and holding the birds between the handler and the dog. This can be a very 'killing' method but causes flushing and trouble with young dogs.

AVAILABILITY

The novice should appreciate that the English Setter Club has a membership which includes the greater part of the breeders of English Setters, but owners are apt to be jealous of their strains and do not sell them unless they feel that the new owners will work them.

14 Pointer

HISTORY AND EVOLUTION OF THE POINTER

The Pointer is a comparatively modern breed of bird-dog evolved in the British Isles during the 18th century and perfected during the 19th century. The original foundation stock was the Spanish Pointer, imported by army personnel either direct from Spain about 1713 after the war of the Spanish Succession or possibly from France in the previous fifty years.

From all available descriptions the Spanish Pointer was a slow, lumbering dog with excessive lip, ear and throat, reminiscent of a Bloodhound, but with a high head carriage when hunting, scenting powers previously unknown in any other breed, and a very highly developed natural quartering instinct. Slow they may have been, but, from all accounts, no game was missed on the ground they traversed. The origin of the Spanish Pointer is obscure though there are various historical indications of its existence, or the existence of a similar breed in Italy in the 13th and 14th centuries. The various French Braque show evidence of origin from the Spanish Pointer.

In the latter part of the 18th century various British landowners and sportsmen experimented with Foxhound, and to a much lesser extent, Greyhound crosses, in an effort to produce a dog with the quartering and game finding ability of the Spanish Pointer, but with increased pace and endurance as a result of the hound cross. The Foxhound cross appears to have been widely used though rarely publicly admitted. Col. Thornton, Thomas Webb Edge, the Lords Derby and Sefton and Mr Antrobus from Cheshire were the best known and most successful breeders of the improved breed, the foundation of the Pointer as we know it.

The breeders who were successful culled all the resultant offspring which showed hound characteristics and gradually, over the years, achieved the type of dog we know today, lithe, clean cut, fast and stylish with the ability to quarter the ground well and hunt for a body scent with the head held high. Unfortunately less intelligent breeders, seeing the success of the cream of the Foxhound cross, repeated the experiment but failed to cull selectively with the result that strains of Pointers came into being, and with further hound infusions, persist to this day, showing the undesirable characteristics of the Foxhound. In appearance, the long ear, the straight face, the round bone, the upright pastern, the cat-foot, the high tail carriage and the coat of two lengths are all evidence of hound type.

Unfortunately some past and present show judges of the Pointer in Britain are prepared not only to overlook these faults, but accept them as contributing towards Pointer type. In working, the undesirable aspects of the hound show in the low head carriage, always trailing for a foot scent, a lack of gay tail action, and a tendency to quarter in circles rather than straight lines. These same dogs will give tongue when chasing, and sometimes even in running, and often show an innate jealousy which makes them virtually impossible to train to 'back'. In view of the fact that jealousy has been bred into both the Foxhound and Greyhound as a virtue, this is hardly surprising.

The first field trial in Britain, held at Southill in Bedfordshire in 1865, and judged on a points or percentage basis, saw Sir Richard Garth's Jill and Mr Fleming's Dandy score 100 per cent, followed by Mr Brockton's Bounce, Mr J. H. Whitehouses' Hamlet and Mr Hanley's Moll with 90 per cent.

The four pillars on which the modern Pointer was built are generally agreed to be Whitehouses' Hamlet, Garth's Drake, Brockton's Bounce and Statter's Major.

J. H. Whitehouse's Hamlet, which made quite a reputation for himself as a field-trialler and a good looker, is immortalised in the George Earl painting of 1868. He seems to have been the first lemon and white to gain public acclaim, though all sources of information mention Mr Lang the gunsmith of Cockspur Street as having had a well established strain of lemon and whites which were top class field dogs, and it is from these bloodlines that Hamlet came.

Hamlet, though small in stature himself, established a line of really good triallers and bench winners, mainly lemon and white in colour. Human nature being what it is, contemporary breeders and subsequent writers set out to prove that the lemon/white colour was undesirable and indicative of a recent Foxhound cross. The original Spaniard had been predominantly liver and white though a few were black/white, and from that point of view lemon and white represented alien blood, though the Pointer of the mid-19th century was so far removed from the Spaniard that the point seems irrelevant. The most important fact is that Mr Whitehouse had a very clear idea of Pointer type and having found it in Hamlet, succeeded in fixing it in his progeny.

Sir Richard Garth, Q.C., owner of Jill, who put up a perfect performance on the judging results of the first field trial, was a great patron of trialling in its early years and fittingly produced one of the great dogs of all time in Drake. Stonehenge describes him as a 'phenomenon among dogs' and his great pace, nose and game handling appear to have set him on a pedestal compared with his contemporaries, Reputedly because of his great speed, he did not stand on point until his seventh season, when the edge of his speed was blunted; instead he flung himself to the ground as soon as he touched the scent of game. From all contemporary accounts he was regarded as an ugly Pointer, a ragged bony dog with a plain head; nevertheless, the painting by George Earl shows him to be a powerfully built piece of machinery with an alert, intelligent head.

Drake was widely used at stud in Sir Richard Garth's ownership and achieved good results, especially on the closely inbred bitches of Lord Sefton's Kennel. When Sir Richard became Chief Justice of India, his Pointers were sold by public auction in 1874 and Drake, by then a seven year old, was knocked down for 150 guineas to Mr Lloyd Price.

Idstone, in 1872, describes Mr Brockton's Bounce as the best dog he ever saw on game. From the available records he appears to have been bred from the liver and white strain of Mr T. W. Edge with the distinctive shadings and markings of that line. A good looking big and powerful dog, his blood is reputedly combined with that of Hamlet in Mr Sam Price's Bang and others from this well known Devonshire Kennel to establish, along with L. J. Bulled, the world famous 'Devonshire' Pointers of the latter end of the 19th and the early 20th centuries.

Statter's Major, very highly thought of in his time, because of his excellent field performances and the reputed authenticity of his breeding, as compared with others, appears to have been of a much heavier type than his contemporaries. Through his sons he has had a very marked influence on the foundation of the Pointer in America.

The end of the 19th century and the beginning of the 20th saw the Pointer at its zenith in Britain. 'The Pointer and his Predecessors' was written by Wm. Arkwright and has rightly taken its place as the Pointerman's bible in every part of the world. The Arkwright dogs have achieved fame to which his book has contributed in some measure, greater or less. His ideas on Pointer type are excellent and clearly expressed, but he became unduly interested in the all-black Pointer, which, using hindsight, would appear to have done harm which it has taken a long time to undo in the various countries to which it was exported.

The all-black Pointer is generally regarded as coming from the Greyhound cross and certainly some of those at present in Europe and the Far East show the tucked up loin and low set tail that could possibly come from the Greyhound. It achieved popularity in Scotland, though it is

57. Field Trial Champion Blackfield Gem, owned by Mrs P. M. Badenach Nicholson, winner of the Champion Stake.

58. Ch. Toberdoney Cilla of Wassettwell, winner of 15 Challenge Certificates and several major trial awards, including one Open Stake. This dog is owned by the author.

possible that the alien blood here was Setter rather than Greyhound. Lord Home, an ancestor of Sir Alex Douglas Home, maintained large kennels of blacks at Coldstream and Lanark and I have heard of others within living memory north of Inverness. Arkwright's original blacks came from Mr Pape of Newcastle, who had earlier based his Kennel on Admiral Mitford's Northumberland blacks. In turn these appear to have emanated from the Kennels of Daniel Lambert of Leicester, as great an authority on fighting cocks as bird dogs, who had to withdraw from the sporting scene in the late 18th century 'because of his great obesity'!

Perhaps the most influential of Arkwright's dogs has proved to be Aldin Fluke. Fluke carried the blood of Hamlet, Drake, Bounce and Major and his blood comes down to the present day through Mr Tom Steadman's Mallwyd dogs to his son David's Maessydds and Arthur Eggleston's Pennines.

Arkwright's contemporary, Dr. Salter, did considerably more for the Pointer than anyone else in his long life time though, apart from in Herr Marr's book on Pointers and Setters, his name is rarely mentioned. Dr. Salter lived for 91 years and judged a trial three years before his death. His Kennels in Essex produced dogs in large numbers. His Pointers were mainly based on the Devonshire breed of L. J. Bulled and Mr Sam Price, owner of Ch. Bang, and if credit can be given to any single person for the foundation of the American working Pointer, it should go to Dr. Salter.

F. C. Lowe, another long-lived Pointer man, whose name lives on in 'Carta Carna' dog food today, owned the Kippen and Bobbing Kennels. At the age of 76 he had the satisfaction of making the most recent Dual Champion (F.T. Ch. and Sh. Ch.) in Ben of Bobbing (1922). Arkwright, Lowe and Salter did much to distribute the Pointer world-wide, dogs from their Kennels going in profusion to the United States, Russia and all over Europe. At the beginning of the present century they shared their fame with the Kennels of J. Lee Bulled, Thomas Steadman, J. T. Eggleston and Isaac Sharpe. J. L. Bulled was the son of L. J. Bulled founder of the Devonshire strain. Doubt is often cast on the purity of breeding of these dogs—some of their contemporaries felt certain that the Foxhound cross had been re-introduced—though this is of little consequence when one realises that recourse has been made to the Foxhound further north in the present century—and also to the English Setter!

Thomas Steadman, a Westmorland man who migrated to Merioneth in his youth, founded the Mallwyd English Setters and Pointers which continued in his son David's Maessydds until 1939. The Mallwyd Pointers are important in the context of the spread of the Pointer in Europe.

J. T. Eggleston, another Westmorian with a variety of sporting interests, bred several good Pointers including Tom Speedy which acquired fame in the U.S. as the sire of one of the all-time great performers Great Island Ringing Bells. He also bred Broomhill Dan who, after export to the United States, sired another household name in trial circles in Becky Broom Hill.

Much of the blood of these dogs, and of the Steadman dogs, was carried by the Pennine Pointers of Arthur Eggleston, son of J. T. Eggleston.

Isaac Sharpe ran a professional training, hiring and breeding Kennel of Setters and Pointers in Banffshire until the last war. Dogs from his Kennel, carrying his 'Stylish' prefix, found their way to all parts of the world and had a marked influence on the breed in U.S. and Europe. The Stylish Kennel dominated the Pointer field trial scene between the first and second wars, either running for Mr Sharpe or in the hands of other owners. Lorna Countess Howe, better known in Labrador circles, built her Banchory Pointers on Stylish dogs with the addition of imported Continental blood through Blackfield Jill, If des Herminettes and Gaff di St. Patrick. The Duke of Montrose maintained the large and successful Isle of Arran Kennel with an infusion of Stylish blood. Mrs Nicolson's Carswell and Miss Prior's of the Boreen Pointers were also successful. Mr Miller and Miss Moir's Tiverton dogs won awards consistently at field trials and on the bench, though their bloodlines have now faded. Col. A. S. Balding's Flashaway dogs and Mrs Mary Holt's Menaifrons are to be found in the pedigrees of many of today's winners.

Since the last war, the working Pointer scene in England and Scotland has been dominated by a few large kennels. The most noteworthy is the Scotney Kennel of the late Lord Rank. Founded on Banchory stock bred from Lady Howe's pre-war importations, blended with some American importations, a steady stream of winners has emerged under the guidance of his kennel manager, the late Wm. Brunt and the excellent handlers Dave Munro and Alf Manners.

Mr Joe Dub, who settled in Northern Ireland during the war, imported dogs from Italy, Denmark, Scandinavia and Germany. He was given the Blackfield prefix, made famous in Europe between the wars by his friend and adviser, the late Waldemar Marr and proceeded to breed a succession of top class field dogs for himself, Lord Rank and other owners. Blackfield Flash, Galla and Airpilot became field trial Champions for Mr Dub himself. Arco, Devil, Fast, Frost, Jet and several others won well for Lord Rank; the beautiful bench winner Blackfield Gem won her field trial title and the Champion Stake for Mrs P. M. B. Nicolson and Blackfield Faust became a field trial Champion for Mr Watney.

In years to come, history should show what a debt the Pointer in this country owes to Mr Dub.

The late Alun Robert's Segontium Kennels played an important part in the evolution of many of today's strains. His F.T. Ch. Segontium Whiskey occurs in the pedigree of most of today's trial dogs and F.T. Ch. Segontium Quill was one of the first of several good dogs in the Cornbury Kennel of Mr Oliver Watney, managed, trained and handled by Arnold White Robinson. The litter brothers Cornbury Viscount and Viceroy each won the Champion Stake for Mr Watney.

 The late Lady Auckland of Cromlix, famous in Labrador and Cocker

Spaniel circles, imported Cam of Cromlix from Count Saladini's Italian Kennel in the 1950s. From Cam she bred two winners of the Champion Stake in Lark and Swift of Cromlix and again won the Champion Stake with Lark's son Swallow of Cromlix, all expertly handled by Angus McLachlan.

In Ireland the post-war situation has been different, in as much as that trials have been supported by many more owners, all operating on a much smaller scale than the large English Kennels. These owners and breeders have used the Blackfield imports and their descendants and other dogs from English Kennels as well as their own imports. Mr John Nash of Limerick who took over the great trialler Int. F.T. Ch. Queen of Glengar on the death of her owner, Mr Barry, has imported from Louis Zavattero's Italian Kennel and also used the Finnish import Paatsan Earl to help to make his Monruad Kennels world famous. Mr Michael Early, Dr. O'Flynn, Mr Seamus O'Donnell, Sean Dennehy, Rev. D. Griffith and a host of others have helped to bring the Irish working Pointer to a height never previously achieved.

With the recent deaths of Lord Rank and Lady Auckland the era of the big English Kennels would appear to be on the wane and present signs are that the gaps will be filled by several enthusiastic small kennel owners, which can only be for the good of the breed. In the 1960s Lady Jean Fforde, daughter of the late Duke of Montrose, revived the Isle of Arran Kennels which successive Dukes of Hamilton had successfully maintained throughout the history of the Pointer. Already, Lady Fforde has made two field trial Champions in Isle of Arran June and Dice and bred another influential dog in F.T. Ch. Scotney Isle of Arran Jack.

Mr R. McElhinney of Belfast has campaigned the litter brothers Scotney Feather and Pompey, bred by the late Lord Rank, with great success at trials in 1973.

TRAINING

The training of a Pointer is a large subject on which many men have many minds. General Hutchinson has dealt with the subject in great depth in his 'Dog Breaking' (1850) and the recent publication 'Training Pointers and Setters for Field Trials' (1973) by Prof. Beazley, Alf Manners and Arnold White Robinson with its helpful diagrams has dealt thoroughly with all aspects of the subject.

My own thinking is that apart from basic obedience there is little that the human handler can teach a Pointer; rather does the handler bring to light the latent instincts and talents of his charge and encourage their growth. Ideally a Pointer from the age of four to five months should have loose exercise over fields twice daily, either on his own or with the rest of the Kennel. Prolonged exercise is not necessary, 20 minutes each time will suffice—though obviously the mature dog will need much more exercise to get him fit before he starts serious work on the moor. By having regular loose exercise the puppy will learn to turn to a whistle or hand signal and to change direction as his handler changes direction: In other words the all-important contact is established between puppy and handler. Without this contact, where the puppy is always conscious of the presence of the handler, further training is impossible. With this in mind, if a puppy runs off out of control at a later stage in training the handler should not pursue him, but should change direction or sit and wait, and the puppy, worried by the loss of his handler, will return.

Having established this contact and having taught him basic obedience, to drop to a whistle and/or hand signal he is by now nine to twelve months old, fit to take out on to the wide open spaces. At first it is best to run him on barren ground, let him get the feel of heather and establish a rhythmical gallop through it. If he is bred on the right lines he should show a natural inclination to quarter his ground at right angles to the wind where he can take advantage of maximum scenting conditions. Good methodical ground treatment is the essential base on which all bird dog work is built—without good ground treatment it is impossible to produce a good dog. Again, if his instincts are true one should see his head lift even higher as he turns at the end of his cast in an effort to assess the conditions that will eventually influence the depth of his cast. Care should be taken at this stage to see that he shows a natural inclination to turn after a reasonable cast; if not he should be turned by a whistle and one should always ensure that he turns into the wind.

With experience, the good dog will learn to adjust his speed and his cast to prevailing conditions. In poor scenting conditions his pace will ease to match his nose. If game is plentiful his cast will be shorter and his turn tighter so that no game is missed. If game is scarce his cast must be wide to be effective and in good conditions the cast will be deeper to compensate for its width.

While I personally regard the Pointer primarily as a grouse dog. I am convinced from experience that the initial introduction to game should be on partridges if possible. The dog which starts its work on partridges will always show the more sensitive nose as a grouse dog and will revert to work on partridges the more easily after a spell of work on grouse. I like to introduce my puppies to the moor in January and February when the grouse are pairing but are still wild and alert. In this way one can see the ability to gallop, quarter and turn and make full use of the wind. The occasional pricking of the ears and forward pull with a high head as he touches a faint scent of a departing pair of grouse, and his interest in the sound of the calling grouse, indicate that the battle is half won and that I have a puppy worth training.

In April, he can be taken on to grassland with the odd scattered pairs of partridges and with a bit of luck one can achieve the first point. The first few points may well be technical flushes, i.e. the birds will fly as the dog starts to come on point, but here patience is required as only by experience will the puppy learn caution and the ability to judge the proximity of game. Under no circumstances should the handler in his excitement rush up to a pointing puppy; nothing is better calculated to send the puppy dashing into his game and chasing off after them. By approaching slowly and by quiet encouragement, confidence and steadiness will be built up gradually.

The pointing puppy should walk up to his game freely; the failure to move in freely, stickiness on point, is sometimes due to his breeding but is more often due to handling and training. If a puppy which flushes game is forced to point by command before he has had ample opportunity to work out for himself the proximity of his game, it is paradoxical to expect him to walk out his game freely. Similarly, if too much fuss is made of the pointing puppy and he is caressed and patted, he may well think that by pointing he has achieved an end in itself. Ideally the trainer should never go closer than three yards to his dog and from that distance walk side by side with the puppy until the game rise. The puppy should drop to the rising game, which is easily taught if one takes advantage of his split second of surprise as the game departs.

Introduction to shot is achieved by different people in different ways. My own preference is that occasional shots should be part of the background kennel noise but I like to have the puppy pointing well and steadily before training him specifically to shot. The first shot on the training ground should be fired by an assistant from far back as game departs after a point, and while the puppy's attention is riveted on the departing birds. Gradually the gun comes closer but never close enough to distract the puppy from his game. I feel that a puppy can be ruined by training it to drop to shot *per se*. I prefer the drop to shot to start as a drop to wing and eventually an association of ideas leads to a drop to shot.

Having got the puppy quartering well and showing the ability to point game, walk it out freely, and drop to it, he can now be run with another dog to ensure that he adds 'steadiness behind' to his other accomplishments. The essence of backing is that as soon as he sees his brace mate on point he himself should automatically adopt a pointing stance. With many of the best Pointer bloodlines, this will happen right away; otherwise he may have to be stopped by a signal and taught to back. I, personally, like my puppies to back naturally with minimal help or interference from the handler, but contrary to current field trial practice, I allow the backing dog to help to walk out the point, ensuring that he never passes the dog who pointed in the first place. I am convinced that aimless and prolonged backing for the sake of good manners, with no participation in the point, can result in the development of a variety of faults including a lack of enterprise in hunting with a tendency to look for opportunities to back, stickiness on point, and in some cases false pointing.

The Pointer has not been bred in this country as a retriever, but purely as a specialist in the location of game. Nevertheless, he is widely used in

America, north and south, Europe and the Far East as a retriever in his capacity as an all-round shooting dog. The vast majority will retrieve successfully and I myself am perfectly happy to allow those dogs which show a natural retrieving bent to complete the job by retrieving fallen game provided they have completed one season purely as a game finder. In this way, their steadiness to game and shot is not allowed to suffer.

CHOICE OF A PUPPY

When deciding to buy a Pointer puppy for work it is sensible to restrict one's choice to those bred from sound working Pointers. The vast majority of Pointers show the ability to point on the scent of gamc, but those bred specifically for show purposes often lack any inclination to hunt, and the ability to point is useless unless backed up by an intense urge to get out and hunt for game.

Temperament is possibly the most important single factor in the choice of a puppy and one which has for too long been neglected in Pointer circles. The Pointer should be a gay, happy, friendly dog with a very strong inbuilt will to please. Without this desire to please, training can become an unnecessarily protracted and wearing exercise. If possible, see the parents and fully developed relations of the puppy in their home surroundings. Avoid those bred from aggressive, discontented, unhappy parents and also those from nervous, neurotic, gun-shy parents. Both faults have been accepted in show and field trial strains on the basis that the pedigree is right and therefore these misfits will breed good stock, but for every one good one bred this way, another nine are produced that are an abomination to all who have to deal with them.

The puppy itself should be bright and alert. An intelligent observant eye, combined with a neat ear, set on high rather than low, and used well, and a busy and vigorous tail action are normally indicative of a suitable temperament. Eye colour will vary according to the colour of the dog; a lightish eye in a liver or lemon puppy will normally darken, though one like a gooseberry, giving a hard expression, will rarely darken to an acceptable shade.

The head should show sufficient stop to give the impression of a dished face. The resultant slight uptilt of the nose simplifies the high head carriage necessary in later life. The nostrils should be wide and the skull sufficiently wide to accommodate the brain and olfactory nerves. The neck should be long, strong, arched and free from excessive throatiness and set well on to a well laid back shoulder. The front legs should be straight and not tied in at the elbow, otherwise he will never gallop freely. The pastern, which acts as the shock absorber, should be long, strong and sloping and not the upright proppy pastern of the hound. The foot should be oval like that of the hare rather than the tightly knit round cat foot which leads to constant trouble in the foot of a galloping dog.

The rib cage should be well sprung and deep to allow for the heart and respiratory functions necessary to provide for both speed and stamina. The coupling should be strong; if it is excessively long it tends to be weak, but if it is as short as has been bred into some strains, it becomes physically impossible for the dog to gallop. The hind quarters should be powerful with a very well turned stifle and well let down hocks. The hind leg, in providing the motive power for galloping, is possibly the most important single factor in the anatomical make up of the dog.

The tail should be of medium length, strong at the root and tapering to a fine point, shaped like the proverbial bee-sting, set on properly, neither too high nor too low and used vigorously and gaily to indicate a happy temperament.

The mature pointer, with his finely chiselled head and hard dense oval bone; with his long well arched neck, well laid shoulder, well sprung rib, strong coupling and well turned stifle should appear as a succession of graceful curves; a classical clean cut lithe active well muscled dog, capable of galloping at speed for long periods. When moving on a lead his front action should show the long reach of the riding horse and not the short high action of the driving horse and his hocks should demonstrate the driving power which will enable him to gallop.

Size, to my mind, is relatively unimportant, provided the dog has power and balance. In the early days of the breed one reads of the small, medium and large dogs; at the beginning of this century the breed was divided into heavyweights and lightweights. My personal preference is for the medium sized active dog; the small dog may lack stamina in heavy going, and the large dog may wear himself out through carrying his own bulk.

Appearance in the Pointer is important. Without type one cannot expect consistent performance, and from a purely personal point of view I can only work my Pointers for three months of the year so I prefer to be surrounded by dogs whose appearance can give me pleasure for the remaining nine months.

15 Gordon Setter

I am very much indebted to my friend, Mrs Mildred K. Adams, Birdham, West Sussex, of the 'Salters' Gordons. Without her help and records I would not have been able to put together this abridged history. I am very conscious of the many gaps and omissions but this is unavoidable where space is limited.

HISTORY

There appears to be no doubt that the Setter was evolved from the old 'Setting Spaniel' and was known in this country before the Pointer. It can be assumed, quoting Dr. Caius writing in 1570, that the original 'Setting Spaniel' was white with big red patches. There were quite a few whole colours, lemon and white, orange and white, and different combinations of colours, but as early as the 17th century 'black and fallow' dogs are mentioned in Markham's 'Hunger's Prevention or The Whole Art of Fowling by Land and Water' as being the 'hardest to endure labour'. In 1776 the Rev. Mr Simons wrote of Setters 'whatever mixtures may have been since made, there were, fifty years ago, two distinct tribes, the Black-tanned and the orange or lemon and white'. The followers of Bonnie Prince Charlie took black and tan Setters back with them to their exile. The Setter was then considered the best dog on grouse whilst the Pointer was preferred on partridges.

No one knows exactly when the 4th Duke, Alexander, of Gordon, commenced breeding at Gordon Castle, Fochabers, but it is certain that he built up a very powerful kennel of working Setters. There were black and tan Setters at Gordon Castle in 1825, as well as black and white, and black, white and tan. Other large estates, mainly in the North of Scotland and the North of England were also breeding Setters, many of them exchanging stock and using the Castle dogs. Among them were Mr Coke of Longford, Capt. Gordon of Cluny, Lord Rosslyn, Admiral Wemyss, Lord Lovat, Mr McNicholls and Mr Brown of Melton Mowbray. The Earl of Cawdor had another strain which he kept as free as possible from English and Irish crosses. They were tri-colour, heavily marked with black and tan. Lord Bolingbroke's Gordons went back to the old strains of

Gordon Castle and Mr Coke's. The black and tans were not exclusive to Gordon Castle although the 4th Duke was reputed to prefer them—facetiously 'the better to poach his neighbours' preserves'. It is said that he introduced a Collie cross, but it seems highly unlikely that one who prized his dogs' 'setting' qualities so highly would jeopardise this; at this time there were many black Pointers and the Field Spaniel was a leggy, settery type of dog, so that it is more likely that he may have used either of those two as an outcross.

The Irish Setter crosses came much later in the breed history and there would appear to be no doubt that it is from this cross that we still have red dogs appearing in the odd litter, and, of course, this cross is responsible for the beautiful rich tan.

I was told by the late George Crawford, who was kennel manager to Prof. Turton Price and who subsequently took over the 'Crombie' prefix on Prof. Turton Price's death, that they, and other big Gordon breeding establishments, always kept a red Gordon in the background to use judiciously when colour began to dull and required improving. I shall detail a pedigree later to show the excessive use made by some of the Irish Setter cross.

The 4th Duke of Gordon died in 1827, and George, the 5th Duke only lived nine years after his succession. During his life the kennel was greatly reduced and on his death a dispersal sale was held at Tattersalls. The top price at this sale was 72 guineas for a four-year-old black, white and tan, Young Regent, by Old Regent out of Ellen, purchased by Lord Chesterfield. On the death of the 5th Duke the title became extinct and the estates passed to his nephew the 5th Duke of Richmond, but in 1875 the subsequent 6th Duke was created Duke of Richmond and Gordon. The Duke of Richmond (5th) had purchased a five-year-old bitch Juno (black and white) for 34 guineas at the dispersal sale and it is generally accepted that he got a strong kennel of working Gordon Setters going again. There were two Gordons exported to the United States in 1842, black, white and tan, either of Gordon Castle strain or direct from the Castle kennels. These were Rake and Rachel. Reports differ on colour after the kennel was revived by the Duke of Richmond, but the principal colour was still black, white and tan, with a few black and tans, but the latter seem to have died out gradually at the Castle. In fact, when Mr Isaac Sharpe of the 'Stylish' kennels bought all the remaining Setters when the Gordon Castle Kennels were closed down in 1907, these were all black, white and tan.

It was the habit of many breeders at that time to name their dogs after themselves. This, naturally, can cause confusion, e.g., the famous Argyle, bred by the Duke of Argyle at Inverary, also Ch. Lang bred by Mr Lang of Clifton, Bristol. This Argyle was the sire of the Rev. T. Pearce's beautiful bitch Regent, Argyle II and Boll, all out of Ruby. Regent appears in all pedigrees many times over.

Other strong kennels from the mid-nineteenth century onwards included

59. Ch. Dandy of Westerdale backing a Swanley Pointer.

that of Mr Josh Jobling, whose dogs were of excellent type and good workers. His dog Ch. Dandie won at the first Sporting Dog Show (Setters and Pointers only) held in 1859 at Newcastle-on-Tyne. He was catalogued at £50. Ch. Moll, bred in 1861, had at least eight owners and finally, at eight years old, was sold at a dispersal sale for £5. This seemed to be fairly prevalent in those days as Ch. Reuben, born in 1863, owned at one time by the Earl of Shrewsbury, changed hands many times. Reuben by Milo (Dandie) out of Lord Rosslyn's Ruin by Grouse (of the McKinnon of Skye strain) ex Duchess, bred by the Duchess of Gordon, was the sire of the famous Ch. Lang whose blood is in all our present-day Gordons. Ch. Lang was priced at the exaggerated sum of £5,000, at the Crystal Palace Show in 1875. Several lines stemming from him found their way to Europe and the U.S.

Mr Lang's kennels also included Argyle II who won 2nd prize at the All Age Setter Stake at Southampton in 1870, and Rhine IV, born 1864, winner on the bench.

The Rev. T. Pearce, Morden Vicarage, Blandford, Dorset, at one time owned the most controversial Ch. Kent who was much sought after as a sire. It is said he served about 60 bitches per year at the height of his popularity and consequently we have Ch. Kent appearing in all our pedigrees. He was undoubtedly very weak behind and it has now been almost certainly established that this was not the result of bad rearing as it is known that he had the best of food as a puppy. Many irate letters appeared in *The Field* on the subject of his unsoundness (also many in praise of him), but he was a consistent winner of the time and undoubtedly sired some brilliant stock, among them Ch. Rex, Young Kent, La Reine, etc., out of Ch. Regent owned by Mr Pearce. He was believed to have come from the Gordon Castle kennels and his pedigree was accepted by the Kennel Club as being by Shot by Jobling's Scamp. Towards the end of Mr Pearce's breeding activities he was of the opinion that Gordons had got too heavy and must be refined at all costs although he maintained 'I have seen better Setters of the black and tan than any other breed'.

Dr. John Salters was given Ch. Rex in 1873 and set about trying to obtain a faster and less cumbersome Setter by using this dog on his own Irish Setter bitch Sal and to several other Irish Setters and English Setter bitches. Ch. Rex had won 1st prize in the Paget Stakes on grouse in 1866, 1st prize Shrewsbury Stakes for All Aged Setters and 1st prize in Talbot

15. Irish Setter.

16. German Short-haired Pointer.
Peter Moxon

17. Weimaraner.
Peter Moxon

18. Hungarian Viszla.

Stakes for Setters on partridges in 1868. Dr. Salters judged the first Dog Show ever held in France, in Paris, 1878, and also the first to be held in Russia in 1898. Judging at St. Petersburg in May, 1900, he had an entry of 11 Gordons in Open Dog.

Mr Parson of Taunton had a very great influence on our present-day Gordons. He bred Ch. Floss, Ch. Beaumont and the beautiful Ch. Belmont, who unfortunately, were exported to Mr Morris of New York—a great loss.

In 1872 the Rev. J. C. Macdonna founded his Gordon kennel. He mated his bitch Bloom (Reuben ex a bitch going back to Kent and Regent) to Stoke's Shot and produced Blossom who soon became a Champion and was exported to Canada. Blossom was mated to several bitches before going to Canada, the most important 'get' being Ch. Marquis.

Towards the end of the century two very famous dogs were Ch. Duke of Edgworth and Ch. Redruth Colonel. These two were great rivals on the bench, always being neck and neck for top honours. Duke was exported to Mr Vandergift of Pittsburgh at quite a fair age.

We now come to Mr Gibbs of the Lorne strain, who exported several Gordons to Russia. He infused a lot of Irish blood into his breeding and,

60. Ch. Scottische of Swanley with the author on Gleneagles Moor.

possibly in consequence of this, owned several dogs in partnership with Mr Robert Chapman, Glenboig, of the famous Heather strain. Mr Gibbs' most notable Gordon was Young Lorne.

The Heather Gordons were exported world-wide, although Mr Chapman did not begin exhibiting until he had, over the years, improved the breed to a standard to satisfy him. In doing this he used a lot of Irish Setter blood to 'fine down' the Gordons. His magnificent dogs had no rivals and he owned the most powerful kennel there has ever been. Unfortunately, he had a disagreement with the Kennel Club over the authenticity of some of his pedigrees and the Heather Gundog Kennel then came to an end. It follows, naturally, that there is always some doubt as to whether some of these pedigrees are correct. His Champions Heather, Grouse, Nap, Crack, Blossom, Ann, Leah, and many others were all in the first flight. Mr Robert Chapman, Jnr., continued breeding Gordons for some years after the kennel was closed and used the prefix Johnstone.

Mr David Baillie, who was for some years manager of the Heather kennels, had some very fine Gordons, among the best being Ch. Garbet Nap and Garbet Crack.

Between 1900 and 1906 Gordons declined very badly numerically and were poorly catered for at shows in consequence. However, the Stylish kennels now came to the fore, and I was told by the late Ernest Scott, kennel manager for many years to Isaac Sharpe, that at one time they had between 300 and 400 dogs—Setters, Pointers, Retrievers and Spaniels—there. They were fed on horse-meat which was driven round the runs by horse and cart, the chunks of meat being thrown in, and it was a case of the survival of the fittest!

The Stylish kennels had the most wonderful combination of workers and show-dogs, the most famous worker being Stylish Ranger, a black and tan, who won the K.C. Derby in 1901, and the K.C. All Aged Stake in 1902. The Champions are, of course, legendary—Billie, Gwendolene, and Scorcher; the latter, a great-grandson of Billie, is possibly the most important out of the Stylish Kennel, his pedigree appearing in a great many of our and the U.S. pedigrees. The Cornwallis dogs are nearly all of Stylish blood.

In 1920 another great name appeared—that of Professor Turton Price, who founded the Crombie kennels (my first two Gordons were Crombie). His first Gordon was Crombie King and was the only champion in the breed in 1922. A great deal of Stylish blood went into the formation of the Crombie strain, and later Prof. Turton Price provided the foundation of several famous kennels, including those of Mr and Mrs Hall (Daven), Miss Reoch (Valiant), Miss Sharp (Dalnaglar) and Mr T. E. Davies Barnlake). Quite a few of the lovely Crombie Gordons, English Setters and Pointers went to the U.S.A., Australia and many parts of Europe.

We have the Painters, the Bydand, the Morar, the Lomond, the Morecroft, the Withinlee, all around the same era, and, of course, the illustrious

Dendy, the Ardale and Daven. A few years prior to the second World War we had Mr A. E. Wright, Winsdon, and Mrs M. K. Adams Salters, who, along with Miss Sadler of Dendy, managed to preserve one or two good Gordons during the war. Unfortunately, in crisis year, 1938, an outstanding dog, Great Scot bred by Mrs Adams was yet another to go to America, to the Blakeen Kennels. The Dellfort, the Kenroe, Highdown and Mrs Rowe's Cairlie (still very much to the fore) were all well known immediate pre-war kennels.

Quite a number of new devotees appeared immediately post-war, among whom we had the Wilson's Pengelly, the Morse's Shippen, the Stanhope's Freebirch, the Misses Huntington's Riggside, Mr Naisby and Mrs Whitwell's Ardagh, Miss Walton's Claytex, the Butterworth's Gaydale, Mrs Lyndesay Smith's Borderland, Mrs Pease's Westerdale, Mr and Mrs Wade's Dawesdale, Mrs van Dam's Speygrounds, Mr Brigg's Petersdale, Mrs Short's Shordale, and my own Swanley. Of these, the Misses Huntington, Mrs Pease and myself also ran our dogs at field trials.

I do not propose to list any kennels more recent than those mentioned. To the lovers of, and newcomers to, the breed who would like to trace back any Gordon pedigree, I would commend to them an extremely long sheet of paper, and Mrs Mildred K. Adams' book 'The Gordon Setter'.

I am often asked as to the suitability of a Gordon as a gundog and/or companion. To both I would unreservedly say yes—have a Gordon. I find them perfect from both angles, easy to rear, sensible and most affectionate. Because of a bad car smash I had to disperse my kennels, keeping only a Pointer and a Labrador, but as the years passed and I was able to walk again, I yearned for a Gordon. My husband always said (having seen the Gordons at shows during a bad patch) 'What on earth do you want a crawling-about beast like that for? No good for anything'. After I got my Gordon and it had been with us for three days the comment was—'great little dog that isn't it?'. The other day he went so far as to say that if he was limited to one dog only it would be a Gordon!

From personal experience I have found that the handsome, affectionate Gordon, has a wonderful nose, no false pointing, and will stand up to the most gruelling day's work and still ask for more. In recent years it has been established that the Gordon can hold his own at trials, the most outstanding dog, by virtue of having won a Challenge Certificate and six Reserves for the Breed, being F.T. Ch. Rowney of Crafnant, owned by Mr G. Burgess. He was by Danny of Bancroft out of Lady of Sarah. The only other Gordon, so far as I know, to come near the long-awaited Dual Champion title in recent years was my own Ch. Scottische of Swanley who had an award of 1st equal at a field trial on Cabrach Moors. She was home-bred by Duke Mearns (Ch. Dalnaglar Baron ex Valiant Pride), out of Highdown Judy of Swanley (sister of Highdown Jayne, dam of F.T. Ch. Highdown Nadine). Lady Auckland's Cromlix Gordons were a great force in the field trial world, winning many awards, but some were

61. Three top Gordon Setters on Cabrach Moor. (Left to right) Duke of Mearns, winner on the bench and field trial awards, with the keeper's daughter; Ch. Scottische of Swanley with Mrs Allan, and Ch. Dandy of Westerdale with the Hon. Margaret Forbes Sempill (twenty challenge certificates).

exceedingly small in build. Miss Macintosh, also, kept the Gordon flag flying well, particularly with her home bred dog F.T. Ch. Boyndie Ruadh, who was a red dog. Other well-known working kennels were Alderbury, Fearn, Wassetfell, Biggin and Mr E. H. Waide's kennel.

We have today the title of Show Champion which I, personally, deplore, but many of the gundog people had a very good pro argument, pointing out that the Kennel Club did not make Terriers pass a qualifier before gaining the title of Champion. Very few Gordon owners, nor indeed any other Setter owners, have attempted to qualify their dogs for the title of Champion during the past few years, but there is now a growing interest in this side from many of the new owners. My husband and I organised a 'Day on the Moors' on grouse during the 1973 season to which we invited Gordon owners. The response was magnificent, 19 Gordons turning up.

We were extremely fortunate in having The Lady Jean Fforde and Mrs Patience Badenach Nicolson there to give their expert advice and instruction. As a direct result of this I feel sure we may look forward to seeing one or two full Champions in the breed before too long. Many owners would be able to give their puppies initial training and later bring them up to at least the standard of qualifying without having the advantage of access to extensive grouse moors or shootings.

The most important thing to remember is that a puppy is just like a young child and must be treated as such. For the first few months of its life, feed little and often, and let it play and sleep.

The selection of a puppy is a difficult undertaking. If for show purposes only, one would, ideally, chose a bold puppy, with a dark eye and a noble (I think describes it) head, not snipey, and, although the Bloodhound cross was used at one time, not showing haw. One would try to get a smooth coat, short back, well-made body, well-turned stifle and short straight tail. Colour and correct markings are also important. If one is wanting a puppy for work only, one would not want a long-cast puppy and would look for well-turned strong hindquarters and, naturally, would take into account the working merits of sire and/or dam.

Good rearing of your puppy is essential to a strong upstanding well-finished dog. I start giving Ostermilk No. 2, brown bread, milk and cod or halibut liver oil between three to four weeks old, then a second meal of scraped meat or minced tripe. Fish also is good. When weaned at six weeks old, the puppies have four meals per day—the Ostermilk, etc., at breakfast time, meat or fish and milk at lunchtime, milk pudding or Farex at tea-time and at night they have hard rusks or some of the good puppy milkwheat biscuits to gnaw, and milk. I also give them a big meaty knuckle bone or similar—this is a great help at teething time.

At around four months old they go on to three meals per day (it is essential that they have plenty of milk, meat or fish and their oil) at this stage I also give a little bone meal. At eight months old they are cut to two meals per day. Although our Pointers and Labradors usually go on to one meal per day at a year old I have found that the Gordons really need their two meals per day for quite a bit longer than this. One must remember that they are not fully mature until about three years old. Comfortable sleeping accommodation and access to fresh water is essential.

TRAINING

Introduce your puppy very gently to a collar and lead in the garden at about three months old, never taking it outside one's own premises until inoculated. At about four months old start giving a little road work, making sure you do not tire your puppy. Increase the distance gradually and you will find there is nothing quite like roadwork to give good feet. It must be realised that, like human beings, there are great differences in temperament and disposition. Some require firm handling, others respond to gentle coaxing, but whichever your puppy may be, always remember to give praise where due and to check when he does wrong. Again, never give a long period of training, a little each day is more effective and, if one begins to lose patience with the puppy, it is imperative to put it back to the kennel or house immediately. Do not start training until five to six months old. Always be consistent in your words of command and decide at the commencement, on the whistle to return and the whistle to drop. I personally use a 'burrel' or 'pea' whistle for return and a single tone for drop.

After the puppy is returning to whistle the next step is to teach it to drop. You will possibly have to press the puppy down for a few days until he realises what you wish him to do at the command drop. Accompany the command with a raised hand and whistle. After a few days, walk backwards away from your dog, counting your steps at the same time making him stay until recalled. Gradually increase the distance and then in time it will be possible for you to drop the puppy on the recall to you and, finally, to walk forward, knowing you can trust your puppy to stay until told to come to you.

Always remember that although one has to be firm in one's training, many good dogs have been ruined by being beaten—a slight slap with a lead and a good scolding is all that is necessary and, with some, not even that.

With your dog now reasonably obedient, the next step is to quarter, which means getting your puppy to range from side to side seeking for game. It may be necessary at first to 'quarter' oneself along with the puppy to get him going properly. Always indicate with your hand the direction in which you are going, invariably running into wind, never down nor side wind, and never allow him to pass behind you when ranging. Having previously taught your puppy to come to whistle it will be easy to keep him within range. There is no experience quite so exhilirating as taking a youngster out on the moor for the first time and seeing the first real 'point' or 'set' on game, after having had the amusement of watching him pointing blackbirds, etc., in the run. As soon as the puppy goes on point get up to him as quickly as possible, slip on a lead and encourage him to go forward until the birds rise, then drop him immediately. 'Drop', 'down' or 'hup' whichever one chooses to use.

If the birds rise before one reaches the puppy (and this often happens) he will almost certainly start to go after them. It is wise then to attach a check cord to the puppy's collar (30 yards or so) and, next time he comes on point, get hold of the end of the cord, gently put a slight strain on it and gradually get up to him. Make him stand for a minute or two on point saying 'steady', then move him in to flush the birds. As soon as the birds get up, drop your dog and then lead him forward several yards away from the scent of that particular covey before starting him off again.

After a few days of this training you will be able to dispense with your check cord and have your dog absolutely steady on point. From the beginning your puppy must be discouraged from chasing hares—this is often one of the hardest things to do, but only time and patience can stop this. 'Backing' is another joy to see. This means that if two dogs are running together and one comes on point, immediately the second dog sees the first on point it must freeze to a point on the spot, thereby acknowledging that the first has found game. All dogs are not natural 'backers' but provided one can take out a youngster with an old hand it can be done by intercepting the young one before it can rush up to steal the old dog's point. To the call of 'steady' and/or hand raised or whistle to drop, the initial obedience training will come into its own. It is helpful to have another person who can work out the old dog's point whilst one keeps the young dog backing until the birds are flushed. I have at all times found the field trial people extremely kind and helpful, keen to encourage
 beginners, and if one has the good fortune to have a dog of that standard

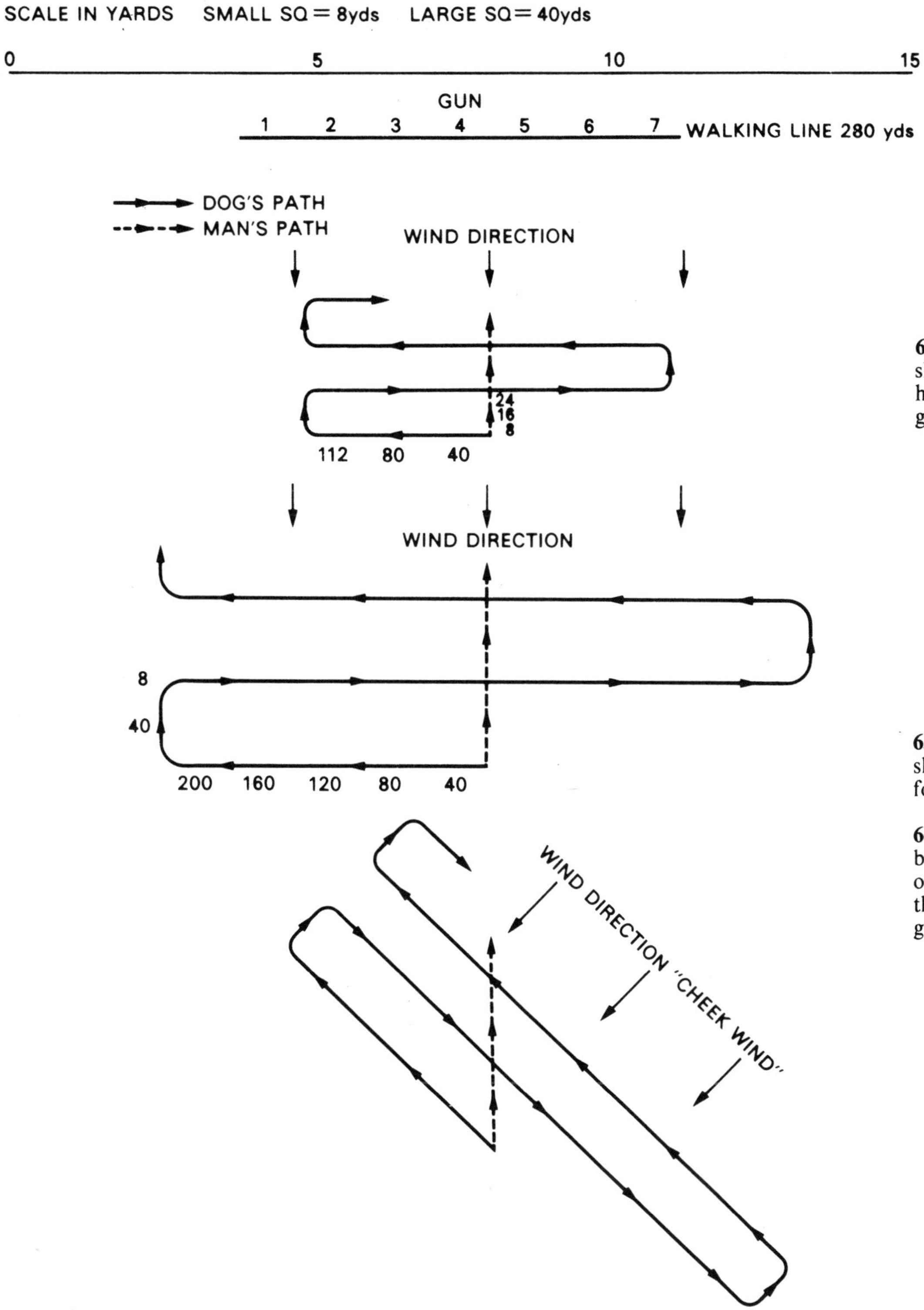

62. This diagram shows very intensive hunting of the ground.

63. This diagram shows a very wide forward cast.

64. The dog is behind or in front of the handler. In these ideal beats no ground is missed.

to compete, they will give every assistance.

Gordons have competed successfully in field trials since 1865. During this century the most outstanding has been Ch. Bouncer of Ardale. Last season, 1973, Mr G. Burgess' Gordons secured several awards. Miss Partridge, and a newcomer to the breed, Dr. Thompson, also featured.

In conclusion I'd like to say that, although a Gordon will retrieve readily and this is done on the Continent, please do not lose sight of the purpose for which they were bred—'setting'. I know one or two of our fairly recent Gordon owners are using them as they would a Spaniel, but there can be no finer sight than your dog on point—if one is lucky—on a sunny day in August on a grouse moor. The scent of the heather, the call of the cock grouse, the bees droning and the odd curlew wheeling away in the distance make all one's troubles and worries fade in the distance and the only thought is how near is the next covey of grouse.

16 Irish Setter

Not surprisingly the Irish setter, with its charming aristocratic temperament, has been gaining in popularity during recent years. This is confirmed by Kennel Club registrations which show it to be the 9th most popular breed in 1972 with 4,782 registrations as compared with 962 in 1965 when it was 30th in popularity.

It is perhaps to be regretted that with the breed originally developed for its working potentialities, the vast majority are now bred entirely for show or as pets, without regard to any ability as gundogs. Whilst these may therefore still have working instincts in them, these instincts have not been developed, tested or used as a basis for selection when breeding.

Anyone wanting an Irish Setter as a gundog would be ill-advised to acquire one without having evidence of the proved ability of its recent forebears, since it is not possible to distinguish by appearance between those with working ability and the others. The good working dog inevitably shows considerable beauty since it could not move with speed and elegance without having correct conformation. Irish Setters are classed as gundogs and, whatever they may be bred for today, they are all descended from the dogs which were evolved and developed as game finding dogs, capable of tireless galloping over the harshest of terrain.

There can be no doubt as to the Irish Setter being a very old breed and, indeed, its origins are lost in the mists of antiquity, if not in the bogs of Ireland. There are references to couchers or setting dogs in the 15th and 16th centuries and these were probably the ancestors of the breeds we know today as setters. Certainly it is only to be expected that well defined breeds did not exist in those days and that there would be many local types of dogs described by the purpose for which they were used rather than by any breed specifications.

Originally it seems that these 'setters' or 'sitters' were favoured for use when netting partridges and quail and were also used for hawking, as indeed they are today. Later the 'setters' were trained as gundogs. Gervase Markham, in 1655, writing in 'Hunger's Prevention, or the Art of Fowling', described their work as follows: 'A setting dogge is a certaine lusty land spannell taught by nature to hunt the partridges before and more than any other chase whatsoever, and that will all eagernesse and fiercenesse, running the fields over so lustily and busily as if there were no limit in his desire and furie; yet so qualified and tempered with art and obedience, that when he is in the greatest and eagerest pursuite, and seems to be most wilde and frantike, that even thus one hem or sound of his master's voyce makes him presently stand, gaze about him, and looke in his master's face, taking all directions from it, whether to procede, stand still, or settle. Nay, even when he has come to the very place where his prey is, and hath, as it were, his nose over it, so that it seems he may take it up at his own pleasure, yet is his temperance and obedience so made and framed by arte that presently, even on a sudden, he either stands still or falles down upon his bellie, without daring once to open his mouth, or make any noise or motion at all, till that his master come unto him and thus proceedes in all things according to his directions and commandments.' Today's handler can want nothing better of his dog.

It is to be noted that the term 'spannell' is used in this quotation, but this does not necessarily mean the same as spaniel does today and probably included the type of dog we now call setters. In this context it is interesting to note that Edward Laverack, the renowned breeder of setters, writing in 1872 states 'the setter is but an improved spaniel'. Without doubt the three setter breeds have much in common.

Rawdon Lee in 'Modern Dogs' published in 1893 refers to various breeders in Ireland who, in the early years of that century, had kennels of red setters though it seems that these were less numerous at that time than the red and white (or white and red) dogs. Some authorities believe that the red dogs were bred from the red and whites, and that the flash of white so often found on the red setter, is an indication of this ancestry. There can be little doubt that the reds were inter-bred with the red and white dogs. The latter may be looked on as a separate, but easily identified strain at a time when there was great variation in the types of dog kept by owners in various parts of the country. Nowadays they are rarely seen although one ran in the 1973 Field Trials.

Setters and Pointers are used to find game in rough cover such as heather, long stubble and young plantations. With the mechanisation of corn harvesting and the intensification of modern agriculture, in contrast to the practices of the 18th and 19th Centuries, the scope for the use of setters on cultivated land has declined, so that today their principal use is on grouse moors. Some moors are termed driving moors, when beaters are used to drive the grouse to the guns waiting in butts, while others are known as dogging moors when the guns follow up behind the setters which seek out and point the game. With the increasing costs of hiring beaters there may be a returning interest in the greater use of setters.

The working setter of today is expected to point game birds and ignore rabbits and hares, but for the individual rough shooter there is no reason why his dog should not be encouraged, if required, to point ground game.

65. Handlers with dogs going forward to compete in a field trial stake. (Mrs Mason with Acornbank Kitreena, left, and Mrs Jarosz with Joanma's Pipit).

Equally, the setter in British field trials, in contrast to those in many continental countries, is not required to retrieve, though many Irish would willingly do so.

The Irish Setter has been described as a dog with an athletic silhouette, robust without heaviness, elegant with an intelligent look, and an assured and noble attitude. When working, an Irish Setter is required to be able to gallop in long heather over rough mountainous country for many hours a day, so stamina is of prime importance. For endurance a strong heart in both the physical and metaphorical sense is required, also good lungs, deep chests, well sprung ribs and four good legs. It is beautiful to see an elegant deep mahogany red dog in full gallop, moving with confidence over heather with a long stride and flowing action. Such a dog will be of compact build, not too large, with a good sloping shoulder, well developed quarters and compact feet, naturally balanced, well muscled and a real joy to behold.

Ability as a gundog, however, goes further, since a good 'nose' is required together with a good brain to use it effectively. It must be of such a temperament that, while full of initiative, it is anxious to please and obey willingly, working for the handler and not for itself. These are the qualities to be admired in the Irish Setter.

Some authorities claim that the Irish Setter does not have as 'long' a nose as some of the other breeds and may therefore get closer to its game before acknowledging scent, but this is very disputable. Certainly, under hot dry conditions the Irish rarely show to full advantage, but in cold wet windy weather they are in their element; indeed, it has been said of the Irish that the worse the weather the better they go.

It is of paramount importance to choose carefully your breeding stock so as to incorporate from both sides the faculties desired in the progeny. Line breeding can be practised with advantages but care must be taken to use only the soundest of stock, both mentally and physically; outcrossing will stimulate the line but the progeny is less predictable. Both conformation and temperament of the working dog should therefore be taken fully into account when breeding, but attention also needs to be given to many finer and less easily defined characteristics to produce top class dogs. These include the way the dog handles its game, bird sense, and the ability to know where game is likely to be found, freezing positively on point, complete steadiness, ease of handling, absence of jealousy, kindness and honesty.

Irish Setters require very good feeding at all times, since with their *joi de vivre* food is burnt up rapidly to provide the energy needed. For strong healthy puppies, feed the dam correctly whilst she is carrying the unborn young and continue to feed her well whilst she is supplying milk for her puppies, since the dam will draw from her own body and skeleton if the necessary food is not readily available from her diet. Many books have been written on feeding and rearing; we all have our own theories. The fact remains that the bitch may be nurturing eight puppies, more or less, which need plenty of protein, carbohydrates, minerals and vitamins, and as her puppies develop she has less and less space to cope with bulky food so she must have more and more of a concentrated diet fed little and often. The bitch should have normal exercise as long as she is prepared to take it, which may be to within a few hours of whelping.

As soon as the puppies are born the beginning of a good relationship with humans can be founded, and this is the start of the making of a trained dog. The puppies cannot see but they can hear and feel, and are sensitive to touch and voice from their first day of life. Treat them with gentle kindness to build up their confidence in man, so enabling them to become the intelligent, great hearted, bold but sensitive dogs required for

66. Thirteen month-old puppy patiently holding her point while waiting for the judges. (Mrs Mason with Baigent Mountain Ringlet).

67. Acornbank Kite on point with Moanruad Ferroda honouring. Dogs distinguishable for judges by coloured cord collars.

working. The introduction of supplementary food should commence at about 10 days old so that the puppies are taking food readily before the mother's milk supply begins to prove inadequate for the increasing demands of the fast growing youngsters. The amount of food given should be increased with the appetite until, by six weeks of age the puppies are completely weaned and the dam only serving as a 'comforter'.

In suitable weather the puppies should be allowed out of their kennel as soon as they are inquisitive enough to want to explore the outside world, gradually extending their range. I get a good deal of knowledge and pleasure watching my puppies exploring a dog paddock which has an area of nearly an acre. By eight weeks of age they know every inch of the ground and will have used their initiative to investigate all smells and sounds, to negotiate obstacles, while learning how to play and come when called.

All training should be kept interesting to produce the dog which will work with the grand Irish style. The work of a setter is to seek and point game birds on a pre-selected area. To do this a dog must quarter his beat correctly, in widths determined by the handler and depths according to wind and weather conditions, galloping at a rate which allows the handler and guns to follow at a reasonable walking pace.

The training for this can start soon after innoculations are completed and I like to get the puppies out into a field and give them their first lesson in quartering as early as 14 weeks of age. Letting the puppy run free and setting off into the wind, it is encouraged to run to the right and is then called back and encouraged to run to the left. The puppy soon learns to turn when called since it is beginning to know its name and is anxious to get back to a familiar person. This exercise can be repeated not more than two or three times on the first occasion and suitable praise must be given. The journey to the field on a lead will have required patience but the return will be easy and thereafter the sight of a lead will be welcomed with enthusiasm and expectations of more work. This little exercise will be repeated at intervals but not too frequently for the puppy to become bored.

68. Preparing to cast the dogs in a brace stake.

Gradually it will learn to run wider than the initial few yards and can be taught to turn to the whistle, always using the same call. Soon the puppy will be taught to sit and wait before being cast away and will then be all the more eager to set off with gusto when it gets the command. Make each outing short and exciting and give praise and encouragement to the puppy.

Dog psychology must be used in training any dog and to produce the first class working Irish Setter the skill of the trainer is needed to develop the full potential of the inherited characteristics in his pupil. The principles of training are similar for all setters but the methods used by individual trainers vary and each dog will require a slightly different approach according to its individual nature. The handler must watch and interpret every movement of his dog just as the dog must learn to interpret the handler's voice, whistle and movements. Dog and handler must establish reciprocal confidence in each other. A large securely wired paddock is invaluable to allow the young dogs plenty of play and muscle building exercise completely free from discipline as a relaxation from training.

When working, commands are given by whistle and hand signals, using separate turning and drop calls. It is essential to encourage the dog to glance regularly at the handler to maintain contact in the partnership. If the handler gets distracted and fails to concentrate on the dog it will inevitably lose contact and probably begin to hunt for itself. As training proceeds the beat is gradually widened to 30 to 40 yards and the drop whistle which at first was only used when the dog is approaching can now be used when the dog is going out, but neither this nor the training sessions should be overdone, making the dog bored or tired. The dog should not be allowed to take too deep a bite of land, since this may cause it to miss birds or prevent the handler from being able to keep up with his dog.

Discipline is essential but must not be harsh or severe to the extent of destroying initiative or you will never get the bold fast running dog that is willing to please, but will have one that in inhibited and never dares to give of his best for fear of doing wrong. The rate of progress will depend on the aptitude of the pupil and ability of the trainer but by nine months the dog should be quartering well and under full control—bitches usually develop earlier than the males.

During training a dog may have shown its interest by pulling up on larks or dropping to the sight of crows, etc., so now is the time to introduce it to game. This should be done in a place with plenty of cover for the birds, whether they be grouse on a heather moor or partridges and pheasants on fields or woodcock or snipe on rough land. Take the dog and set him away with the wind in your face and blowing straight into the dog's nostrils. When there is a scent of game the dog will break his beat and take an extra short cast to quest for it, before hesitating and pulling up. The handler should then go quietly up to the dog, slip the lead on, talk to it gently to encourage it to concentrate on the scent and thereby help it to freeze. After a few moments ask the dog to move forward and let it take you quietly, step by step to the game. Immediately the bird flies up, drop the dog and praise it again—the lead is only there to prevent an impulsive flush or chase. Then let the dog smell the ground where the bird was and take a few short casts on the lead to see if any birds are left. The first point may not have been very spectacular but is enough for one day for the dog to go home to reflect on his achievements.

The dog can now be given more opportunities to find game, but one or two points is enough on any outing. As soon as it can be trusted, after a point, to walk up the birds kindly without fear of a chase, the lead can be omitted. Next the dog must be taught to ignore rabbits, which is easier at this stage since it will have less interest in fur once it has had bird scent. Hares may be pointed, but if they are on the move the dog must ignore them and never follow hare lines.

At any time now the dog can be introduced to the noise of a gun. Let the dog find game and proceed as before (with the lead on) until birds get up, then have the gun fired at a distance from the dog, dropping him to shot at the same time. Most young dogs will never be concerned if the sound of a gun is introduced in this way, but a shot too close to the dog, and for which it can see no reason, can cause trouble. At a later date a bird can be shot over the dog and it is then often advisable on the first occasion to have a lead on to prevent the dog from running in. The dog must always drop when birds get up so that the guns can shoot without concern for the dog; it also enables the guns to reload before any further birds are flushed.

Should the dog refuse to pull up on game scent or deliberately, after winding birds, rushes to flush and possibly to chase, it can be put on a long, light and smooth cord attached to a collar. Working the dog as before, the cord will pass from side to side in front of the handler so that the dog can be stopped when it scents game. A quiet pip on the whistle gains contact with the dog which is held on point while the handler walks slowly up the cord to the animal. The dog must be physically strong, galloping well and working his land with enthusiasm before a cord is ever used otherwise it will tend to inhibit him.

The dog now has an elementary knowledge of game finding; the rest should come with experience and the correct handling for that individual dog according to his style and character. Some dogs, after coming on point, tend to crouch and this could be an inherited trait from the days when they were used for netting, hence the term croucher or sitter. Don't presume that once a dog has pointed game that it will never miss birds, run into them, or do things that you cannot understand. Dogs can have off days and the weather be contrary, winds changing from no wind to gales, or shifting winds going full circle while the dogs are running, days when everything is so hot and dry that no scent is held while the pollen from the heather raised by the dog looks like the vapour stream from a jet plane. Game is also unpredictable, being wild one day, getting up well before a dog can point them while, perhaps, the next day sitting so tight that they almost refuse to fly. Dogs may be able to point game a hundred yards or so away one day and yet only a matter of feet another.

Some dogs show their best work in competition while others, which are grand shooting dogs, seem unable to cope with the added tension of field trials. This is manifested in various forms, such as being jealous, pinching points, or leaving the game finding to the other dog.

The main group of field trials are run in the North of England and in Scotland prior to grouse shooting, while a few other field trials are held in the spring and autumn on partridges and pheasants, in each case at the kind invitation of owners and shooting tenants. The trials are run by various clubs under the overall Kennel Club rules. Puppy Stakes are for those born during the previous or present year, Novice Stakes are usually for non-winners, while Open and All Aged Stakes may, like the Puppy and Novice Stakes, be for all Setters and Pointers or confined to single breeds.

To make a dog into a Field Trial Champion it must win two Open or All Aged Stakes with eight or more dogs competing, one of which must be a Mixed Stake; alternatively, winning the Champion Stake gives it the title.

Field trials are now very popular, with even 50 entries in Open Stakes as compared with a few years ago when it was impossible to run some breed stakes. Various methods are being tried to restrict the entries to a reasonable number thus enabling the two Judges to see sufficient of each dog working to make their appraisal and awards. Lots are drawn and dogs are run in pairs, not as a knock-out since both dogs may be good enough to go into the following round. The Judges look for style, pace,

69. Brace work showing handler having gone forward to Acornbank Peregrine on point, leaving A. Kitreena backing.

quartering, decisive pointing, bird handling technique, backing ability and general quality of work. Two or three rounds are usually run, discarding the dogs which make mistakes, and usually running the best dogs together for final placings.

Grouse shooting starts on the glorious 12th (of August) whatever the weather. Bird dogs and picking-up dogs are bundled into Land-Rovers cheek by jowl with guns, keepers, handlers, refreshment and protective clothing, before setting off on the track, however rough, to the particular piece of moor chosen for that day. About four guns is the ideal number to shoot over dogs, using one dog at a time in turn throughout the day. Three or four setters will be needed and those not running are walked on leads by a keeper. It is now that the dog will prove its worth, cutting out a wide beat, ignoring hares, sheep and deer, hunting the often thinly populated grouse moor, diligently, tirelessly and unerringly, knowing there is no other dog pushing hlm into a fast competitive gallop but only he to go at the pace that suits that day's conditions.

Coming on point perhaps high up on the mountainside or even out of sight, waiting on point whilst the handler and guns struggle through knee-deep heather and rocky terrain to reach his side, may take 15, 20 minutes of more, and only after the guns have clicked their safety catches off will the dog flick his eyes to the handler for permission to go forward—all this will have been done without one word or whistle from the handler. To see an Irish working thus is to see him at his best—no sight on this earth can give more pleasure.

Some idea of the different situations that arise during field trials and/or shooting can be seen from the following expressions which are commonly used.

Cast: this is setting a dog away to work a 'beat'.

Forward cast: refers to the dog in front when a brace is working.

Quartering: the uniform covering of a beat from side to side of the handler.

Your Beat: is the width of the area to be 'quartered'.

In the line: the line of guns and/or judges.

The Gun: is the man who carries the gun.

Your line: is the centre line of a beat and therefore at right angles to 'The line'.

Drawing: a dog which has got a slight whiff of game carefully moving forward to better his position before 'Pointing'.

On Point or Pointing: a dog which has stopped because it has scented game birds ahead.

Frozen on Point: a dog absolutely immobile pointing game.

Steady on Point: continuing to stand firm on point.

Claiming a Point: is the raising of one hand by the competitor to confirm to the Judge that his dog is 'on point'.

Walking up: when dog and handler move up to birds following a 'point'.

Blinking a Point: is done when a dog which has been firmly pointing, just leavcs, even though birds are there.

Work out Point: when birds have been put up following 'a point' and the dog is then asked to work on to find any remaining birds left from the covey.

Sticky on Point: when a dog 'on point' refuses to move forward on command.

False Pointing: repeated pointing where there is, or has been, no game.

Backing: is the term used when the 2nd dog of a brace stands 'on point' behind the 1st dog.

Honouring a Point: is the term used when the 2nd dog of a brace stops because it sees the 1st dog on point.

Mirror Pointing: when both dogs in a brace simultaneously think the other is 'on point' and 'honour' each other.

Too big a bite: this is when the dog goes too far forward on each cast or turn.

Boring Out: when a dog goes straight out ahead and does not quarter its ground.

Turns wrong to the Wind: a downward turn at the end of a cast.

Spanieling: when a dog takes slow short casts instead of working wide and freely.

A Flush: is when a dog runs into birds and puts them to flight without him having acknowledged their presence.

Down wind flush: when a dog comes down wind into birds which it has not had an opportunity of scenting. This is not necessarily a fault.

Wild Birds: not to be confused with a 'Flush' since 'wild birds' are ones which get up of their own accord before the dog has had an opportunity to point them.

Dropping to shot and wing: when birds get up and/or a shot is fired the dog is required to go flat down.

Creeping: a dog having come 'on point' but then without authority creeps forward, possibly 'flushing' the game.

Missed Birds: birds put up from ground which has been worked over by a dog.

Chopped Bird: one killed by a dog.

 Chase: when a dog chases fur or feather. Chasing sheep is a cardinal sin.

Hare Lining: when a dog drops its nose and follows a line previously taken by a hare.

Noisy Handler: a handler who shouts or uses a whistle excessively.

The credit for these wonderful dogs must be given to our predecessors but it is our duty to continue to the best of our ability to breed only those which will preserve that which they have founded.

Many names of famous breeders and dogs of bygone days have been listed elsewhere and are therefore not repeated here, but it is appropriate to note all those which have won the accolade of Field Trial Champion, recognising nevertheless that these awards depend on the handling as well as the dog, and that many wonderful dogs may never have been entered for trials.

FIELD TRIAL CHAMPION IRISH SETTERS IN THE UNITED KINGDOM SINCE THE TITLE WAS CREATED IN 1899

Compiled by Colonel A. S. Balding, M.C.

1899	G. H. Cheetham's	*Honeysuckle*	(b)
1913	W. Wilson's	*Cloghan Norah*	(b)
1924	R. O'Kelly's	*June*	(b)
1927	Mrs F. Nagle's	*Sulhamstead Sheilin D'Or*	(b)
1929	W. J. Patterson's	*Rhu Gorse*	(d)
1929	Mrs F. Nagle's	*Sulhamstead Valla D'Or*	(b)
1930	W. A. Bowen's	*Miss Chipps*	(b)
1931	Mrs B. E. Yeoward's	*Cymwran Barberry*	(b)
1932	Mrs F. Nagle's	*Sulhamstead Baffle D'Or*	(d)
1932	Mrs F. Nagle's	*Sulhamstead Token D'Or*	(b)
1932	Mrs F. Nagle's	*Cloghan Belle*	(b)
1933	Mrs F. Nagle's	*Sulhamstead Dinah D'Or*	(b)
1934	Mrs F. Nagle's	*Sulhamstead Bluff D'Or*	(d)
1936	Mrs F. Nagle's	*Sulhamstead Ban D'Or*	(d)
1936	Mrs F. Nagle's	*Sulhamstead Brantome D'Or*	(d)
1937	W. A. Bowen's	*Rowen of Kilpeacon*	(d)
1938	Mrs M. Holt's	*Menaifron Keane O'Moy*	(d)
1938	Mrs M. Holt's	*Menaifron Rosina O'Moy*	(b)
1939	A. L. Spier's	*Garry of Burtown*	(d)
1950	Mrs F. Nagle's & Miss M. Clark's	*Sulhamstead Baron D'Or*	(d)
1950	Rev. J. McMenamin's	*Tremont Stut*	(b)
1952	W. Kehoe's	*Juliet of Burtown*	(b)
1953	T. R. G. Patterson's	*Cooel Lassie*	(b)
1954	W. T. McMenamin's	*Red Admiral of Rye*	(d)
1954	Mrs F. Nagle's & Miss M. Clark's	*Sulhamstead Basil D'Or*	(d)
1954	Mrs F. Nagle's & Miss M. Clark's	*Sulhamstead Bounce D'Or*	(d)
1955	D. McMenamin's	*Roscommon Red Grouse*	(d)
1955	G. Spence's	*Red Bloom*	(b)

1956	Mrs F. Nagle's & Miss M. Clark's	*Sulhamstead Bo Peep*	(b)
1957	Mrs F. Nagle's & Miss M. Clark's	*Sulhamstead Nimbus D'Or*	(d)
1958	J. Nash's	*New Square Red Lassie*	(b)
1959	A. L. Spier's	*Sherry of Curracloe*	(b)
1959	Mrs F. Nagle's & Miss M. Clark's	*Sulhamstead Nina D'Or*	(b)
1961	Mrs F. Nagle's & Miss M. Clark's	*Sulhamstead Nearco D'Or*	(d)
1961	Mrs F. Nagle's & Miss M. Clark's	*Sulhamstead Nibs D'Or*	(d)
1961	P. Bradley's	*Moylrath Blaze*	(b)
1961	A. C. Moore's	*Tullyelmer Lark*	(b)
1962	P. F. Goodwin's	*Rahard Belle*	(b)
1964	A. S. O'Connor's	*Ballymac Eagle*	(d)
1964	R. A. Stewart's	*Slievebawn Minny*	(b)
1968	J. M. Beazley's	*Ridgetor Firbanks*	(d)
1968	J. Nash's	*Patricia of Killone*	(b)
1970	J. Nash's	*Bena of Maytown*	(b)
1970	Miss M. C. Aitkin's	*Danske Moymore Lulu*	(b)
1971	K. Patterson's	*Divis Gem*	(b)
1972	R. McElhinney's	*Cleo of Maytown*	(d)
1972	R. McElhinney's	*Grouse of Maytown*	(d)

No account of working Irish Setters would be complete without special reference to the Sulhamstead kennel of Mrs Nagle, in which Miss Clark was later a partner, and for which George Abbott was the trainer. The combination of Mrs Nagle's dogs and George Abbott's handling started with Sheilin D'or, the first Irish Setter he trained, which was placed each time she ran in the Champion stake, winning it in both 1927 and 1930 and being 3rd and 4th in the other years. From then on, until George Abbott, one of the greatest trainers, retired in the mid-1960s and the Sulhamstead setters were no longer entered for field trials, this kennel produced champion after champion to a total of 17, and surely this must have been the greatest kennel of Irish Setters in the world.

Another outstanding kennel, now at its zenith, is the Moanruad kennel owned by John Nash in Ireland. It was started in 1946 with a background of famous dogs and bitches, combined with a knowledge of training and handling inherited and acquired from the great masters. His original Red Blaze line produced international field trial champions, Rahard Belle, Moanruad Chilly Breeze, Moylrath Blaze and many others. New Square Red Lassie, bred by Dick Barry, was his first field trial champion and probably the greatest he has owned, but something good was to be expected with field trial champion Portlairge Blacksmith as her sire and the legendary field trial champion Dunboy Rock as her grandsire. This bitch produced Moanruad Ambassador, rightly named as he had such influence on working Irish Setters throughout North America and became a dual champion in Canada. His sire was international field trial champion Derrycarne Red Admiral of Rye, who had a noted influence on the breed.

Another outstanding bitch from this kennel was international field trial champion Una Moanruad who won a 'grande quette' trial in Europe, the first Irish to do so for many many years, beating that great pointer dog, Elu of Eutom.

Another prepotent sire was the Rev. A. S. O'Connor's international field trial champion Ballymac Eagle, whose progeny over several generations are now making their mark with many successes in field trials. Ballymac Eagle, who was bred by John Dennehy in Co. Kerry, had the misfortune when a year old to break a shoulder, and yet won the Irish Red Setter Stake at 2½. Among other achievements, he won the Irish Championship in 1964, was 2nd in 1965 and 4th in the Champion Stake in Scotland in the following year. International field trial champion Bena of Maytown, a daughter, won the Irish Championship in 1970 while another daughter, Acornbank Peregrine, was 3rd in the Champion Stake that year. Grouse of Maytown, a son, won the Champion Stake in 1972, while Acornbank Kitreena was in the Champion Stake awards in 1973, and these and other descendants, including field trial champions, have had many notable successes in other field trials.

Another great dog, leaving his mark, but from entirely different lines, is Tom Kennedy's field trial champion Punch Bowl, winner of the Irish Championship in 1962.

When we study the extended pedigrees of our dogs, we need to think of the distinguished names and breeders over the centuries who have done so much to establish the quality of the working Irish Setter of today. We, who with humility aim to emulate the success of the classical dogs, will continue to breed only those which preserve that which they have founded, so that future owners and trainers will be able to watch with admiration great-hearted Irish Setters and, in their turn, thank those who have kept faith with this magnificent breed of dogs, full of beauty, stamina, courage and brains.

IRISH FIELD TRIAL CHAMPION IRISH SETTERS

Compiled by Mr John Nash

1925	Major R. O'Kelly's	*June*
1926	Major R. O'Kelly's	*Gilpin*
1928	G. Bennett's	*Eilteog*
1930	Major R. O'Kelly's	*Heathericks Griar*
1932	Colonel Bowen's	*Ann of Aileach*
1936	Mrs M. Holt's	*Menaifron Keane O Moy*
1937	Marchioness of Waterford's	*Curraghmore Biddy*
1938	A. L. Spier's	*Garry of Burtown*
1940	Dr. W. M. O'Sullivan's	*Dunboy Rock*
1941	P. B. Galwey-Foley's	*Red Skin of Qu'appelle*

1946 Mrs Robinson's & P. B. Galwey-Foley's	*Tremont Dreamland of Qu'appelle*
1947 Mrs Robinson's & P. B. Galwey-Foley's	*Tremont Stut*
1948 R. W. Barry's	*Red Spinner*
1952 P. J. McNamara's	*The Blacksmith*
1952 W. T. McMenamin's	*Derrycarne Red Admiral of Rye*
1953 A. L. Spire's	*Rena of Burtown*
1953 W. Kehoe's	*Juliet of Burtown*
1956 P. F. Goodwin's	*Termanagh St Rua*
1956 J. Nash's	*New Square Red Lassie*
1957 D. McMenamin's	*Roscomrobe Red Grouse*
1958 D. V. FitzGerald's	*Prince of Kilmurray*
1958 J. Nash's	*Moanruad Chilly Breeze*
1958 A. L. Spire's	*'Sherry' of Curracloe*
1959 P. F. Goodwin's	*Rahard Belle*
1960 T. J. Kennedy's	*Punch Bowl*
1961 P. Murnane's	*Moanruad Admiral*
1963 R. P. Stakelum's	*Rusty of Televara*
1964 J. Nash's	*Una Moanruad*
1964 Fr. L. Marron's	*Latton Lass*
1964 S. Dennehy's	*Ballymac Eagle*
1965 Dr. D. Hanly's	*Cahin Miss Chips*
1965 R. A. Stewart's	*Slievebawn Minnie*
1965 R. P. Stakelum's	*Sherry Bowl*
1966 J. Nash's	*Moanruad Dan*
1968 P. Bradley's	*Moylrath Blaze*
1968 Fr. L. Marron's	*Latton Blarney*
1968 J. Nash's	*Patricia of Killone*
1969 Rev. R. Dinneen's	*Galtee Grouse*
1969 J. Nash's	*Bena of Maytown*
1969 J. Nash's	*Moanruad Latton Jewel*
1970 S. Dennehy's	*Knockmore's Mona*
1970 J. Nash's	*Moanruad Sprite*
1970 E. J. Cunningham's	*Red Revolution of Fallows*
1970 Mrs E. G. Davidson's	*Bronder of Clontarf*
1971 J. Nash's	*Moanruad Kaye*
1971 J. Nash's	*Knockmore Red Molly*
1972 R. McElhinney's	*Cleo of Maytown*
1973 R. P. Stakelum's	*Hopeful Betsy*

17 German Shorthaired Pointer

HISTORY

This breed was first registered in the German Kennel Stud Book in 1872. Hektor I was given No. 1 in Volume I. He was the outcome of fairly intensive and selective breeding. In the 17th century the Germans imported the Spanish pointer for hunting as a bird dog. This did not fulfil their needs as they also wanted a dog which would trail wounded game. For this purpose they crossed the pointer with the Hanoverian Schweisshund, a dog similar to our bloodhound. This produced a heavy dog which was capable of the work required and was called the 'German Pointer'.

By the 19th century, when shooting had become a sport as opposed to a means of acquiring food, consideration was given to beauty as well as ability. Field Trials became popular and the English pointer was much admired both for its speed, nose and elegance. The English pointer had been bred by crossing the Spanish pointer with the foxhound. This smaller, faster dog was cross bred with the German pointer and, in 1872, was accepted and registered as the German shorthaired pointer.

The German name for the breed is the Deutsch Kurzhaar and the all-governing motto behind the breeding of this dog was 'through efficiency to type'. This indicates very strongly how the importance of working ability controlled the breeding. The type is the culmination of the physical and mental qualities necessary in such a dog. A volatile dog full of speed and enthusiasm is necessary for the air scenting pointer—speed is necessary but it is useless without purpose. Equally, if the sense of smell is blunted then speed is useless. Scent will stimulate the work of a dog providing it has the nose to recognise and the intelligence to acknowledge it. For tracking and trailing both live and dead game, a calm and persevering nature is necessary and it is well known that for such work there is a need for a far more sensitive nose than for air scenting. This indicates that one of the most important factors in the make of the breed had to be the ability to adapt and to adjust to the existing conditions at any one time. The correct use of all the inbred instincts was vital to achieve success.

The appearance of the G.S.P. should be that of a compact, medium-sized dog, standing over plenty of ground. The expression shows purpose, friendliness and kindness, while power and perseverence should be indicated in the conformation. The Standard for dogs is 23/25 inches at the shoulder and 55/75 pounds in weight; for bitches between 45/60 pounds and standing from 21/23 inches. A variation either way is allowable as proportion is more important than exact weight and size.

The head is clean cut with no definite stop on the nasal bone as there is in the English pointer, the ears are moderately long and rounded reaching to the corner of the mouth, the jaws powerful, long and square enough to carry game weighing up to 20 pounds; teeth should fit in a scissor bite. The eyes are almond-shape in varying shades of brown to tone with the colour of the coat; a light eye is very undesirable. The neck is strong and muscular, shoulders long and sloping, chest deep, back short and hindquarters powerful. Feet are compact and tight with stifles well bent, tail is set high but carried down or horizontally when working—a gay tail is objectionable. The coat is short and coarse. In all, grace and power should give the appearance of a thoroughbred. All the points mentioned are not Show points, they are points necessary in the make-up of a working dog whose job is to cover a variety of ground at varying speeds with sufficient stamina not to tire.

The tail is docked leaving approximately two-fifths of the original length. Many people query the reason or necessity for this. This is two-fold. The most important probably is to avoid damage when the dog is working in thick cover. It might be argued that the English pointer is not docked, but the reason is obvious, for he is not expected to work cover. If a tail is damaged, healing is a long and slow business, for unless the wound is covered it will continually be exposed to further damage—and who wants to go shooting with a dog with a bandage round its tail, if indeed it can be kept on!

The second reason is also logical. The dog uses its tail as a ship its rudder, for the purpose of changing direction, and by curtailing the length it was considered that hindquarter musculation would have to be more

70. Ch. Littlestat Condiment pointing partridges.

highly developed. Plenty of altercation has taken place concerning the docking of tails. A few shooting men have, in fact, left the tails undocked and have had no trouble, but the vast majority would not under any circumstances consider it practicable to leave the tail undocked. I have owned only one dog with a long tail. He came to me at the age of eight, so I had no choice, but during the shooting season it was a constant source of trouble. By the end of the summer it had healed up only to start again once shooting began. Docking for beauty is totally unacceptable to me, but not to dock where it is necessary for the comfort of the dog is sheer cruelty.

I also feel strongly about dew claws. The removal of these is not mentioned in the Standard of the breed, nevertheless the breeders in this country who breed for work, remove dew claws. I do hear the argument that by leaving them on it makes climbing up banks, etc., easier. I find this difficult to believe when I see my own dogs coping just as adequately without this aid. The other side of the picture shows when, by catching these unwanted claws on some obstacle, damage is done to them. This causes intense pain, and often necessitates an anaesthetised operation to remove the entire toe.

Colour leaves vast scope for variation. Basically G.S.P.s are either liver and white, or black and white. They can either be ticked or flecked. Some models are entirely liver with no flecking, others can have a lot of white with only one or two small liver 'plates', as the Germans call them. Some are ticked all over giving the effect of a silver colour, though pepper and salt might be a better description, but in nearly every case the head is liver with perhaps a white flash down the centre of the forehead, or a white star. In some cases the white is very white giving a beautiful if flashy appearance, others just look grubby all over.

Shooting men have various views on markings, some ask for a dark coloured puppy, others want plenty of white so it can be seen either in the heather or wildfowling. Then you get another man who says that the dog which stood out best in heather for him was all liver coloured, and another who affirms that if he can see his dog wildfowling, so can the duck! I just give up and try to provide what they ask for without entering into any discussion. The only strong views I have are that when docking tails I always try to leave a small white tip even if it means a slightly longer dock being necessary. With this it is always possible to spot your dog in the distance and makes it easier to find it on point in cover. I love a flashy looking dog, just as I love a bright chestnut horse, but it's not the colour that makes the character, and it's character I'm looking for when choosing a puppy.

Now we come to the somewhat controversial black and white G.S.P.s. Here again you can get them almost entirely black or a white flecked dog with only the odd black patch. My first black bitch was black all over except for a small flecked mark on her chest. I think I must say here that,

71. 'To chase or not to chase?' – A G.S.P.
Peter Moxon

to me, the epitome of the markings of a G.S.P. is the flecking or ticking—a self-coloured dog, be it black or liver, is not really typical of the breed, but that is only my opinion.

The history of the black and white G.S.P.s is interesting. In the late 19th century, because eyes were getting too light, a black and white English pointer was imported and bred to a liver and white G.S.P. This produced some black pups and some liver. The blacks, or Prussian Shorthairs as they were called, had a separate register once they were established. They were not popular in Germany and were used only to improve pigmentation by breeding from them about once every five generations. The Swiss and Italians were greatly in favour of them and our own line in England can be traced back through the Swiss blood to Wach Luna I, the first to be registered in the German P.R.K. Stud book. Apart from a separate Stud book, the Standard was separate. This is important as can be seen by the fact that the Americans first imported G.S.P.s into the U.S.A. after the First World War. They imported liver and white ones and accepted their Standard which did not include the black and white. In 1937 the Germans amalgamated the Standards so that either colours were registered in the same Stud book. So, when the first G.S.P.s came to this country after the Second World War, the Standard in Germany covered both blacks and livers under the heading 'colour'. This is the Standard accepted in England—the same as that on the Continent but not in America. I hope the Americans will see fit to alter this situation in the future, as, in my opinion, it is definitely to their loss.

Until recently there have been very few black G.S.P.s in this country. In 1963 I acquired my first black bitch. Her sire had been imported from Germany and mated to a liver bitch here. She proved a good worker and I sold her to a policeman in Ulster on the understanding I should have a black pup from her first litter. This produced Zombi who arrived from Ireland at 10 weeks. I cannot speak too highly of this dog. He is black, I admit, and I don't maintain that the blacks are necessarily better than liver G.S.P.s but my black dog, to me, is outstanding for the following reasons. Firstly his nature is everything any shooting man could ask for; he is kind, both to people and all other dogs. He will put up with puppies hanging on his ears, other stud dogs measuring up to him in an aggressive fashion and even a fighting dog, which he will answer back with as good as he gets and forget all about it afterwards, holding no malice. Nor will he attack a similar looking dog next time he meets one. Water, to him, is as necessary as air, and he will swim for hours for the sheer pleasure of it. His retrieving is reliable and intelligent; he will sit quietly on a drive marking every bird shot by the guns he is serving, but should one be winged or otherwise wounded he has disappeared before I am aware that it was a runner and he seldom returns without it. His pointing ability has always been remarkable in that from a tiny puppy he has never shown any wildness to flush. He will come on point staunchly and remain so until approached, then very quietly will stalk forward until within range to flush the game. He does this with the minimum of excitement and then stands rigidly marking the flight of the bird or birds, or the fluttering of the squeakers, making no effort to interfere with them whatsoever. I have never taught him this, and, in fact, in training have found it a time-consuming lesson for other dogs.

All these attributes appear to be dominant in his make-up so that the majority of his progeny carry them. Conformation-wise he lacks the elegance of the modern G.S.P. and more resembles the heavy German dog of some years ago, but if mated to a finer bitch he has produced some very good looking youngsters. There are still not enough black G.S.P.s in the country to mate black to black without in-breeding, on which I am not keen, so a black is mated to a liver, and the result is that some of the pups are liver and white and others black and white, never tricolour. Of the nine wives to date that Zombi has mated there have been 41 liver pups and 40 blacks. I think I can safely say that in England today a shooting man is just as likely to ask for a black pup as he is for a liver one, as the blacks have proved themselves to be good workers over the last ten years.

As you see from the history of the breed, the G.S.P. was first used as a bird dog, then tracking and swimming was included in his work. The Germans needed not a dual purpose dog but an *all* purpose dog. They used such dogs for pointing and retrieving feathered game of many varieties including duck and geese, and they were used for rabbits and hares and also for hunting deer. In Germany today the field trials test all these qualities. A very high standard is required of their Grade Ia dog, both in pointing and retrieving, and also tracking and stamina. When judging these trials use of nose, staunchness of point, backing, use of the wind, obedience without interference of initiative and keen willingness for work are the requirements for a winner.

Since the breed was introduced into England after the last war, great strides have been made to understand the working ability and potential of this type of gundog. A very active Club was inaugurated which helped with training and breeding, started field trials, and eventually acquired Championship Status for these trials. The schedule of a trial requires the dogs, which are run singly under two judges, to quarter the ground in search of game, to point game, both fur and feather, staunchly, to be steady to flush and shot and to retrieve tenderly on command, both from land and water. The Trials take place on land kindly offered by friendly estate owners, on live game, and the day is conducted as closely to a shooting day as is ever possible at a Field Trial. The judges are looking for the best all-round shooting dog, and for this reason a marking system is used to simplify the difficulty of so large a task. It is very easy to be influenced by an outstandingly classic point or an excellent retrieve on a difficult runner, but what use is this dog if he does not complete the water retrieve, which in the Open Stakes is a blind retrieve of a bird from across water with no shot fired, a cold blooded but practical exercise for any useful gundog. The winner must have acquitted himself well on all counts; in fact, at the end of the day I reckon its owner really has got something of which to be proud.

Further explanation here can be given of the marking system obligatorily used by all Judges who are members of the German Shorthaired Pointer Club. Up to 10 points can be given in each of the nine categories: 1, Ground treatment; 2, Pointing; 3, Retrieving; 4, Game finding; 5, Steadiness; 6, Facing cover; 7, Style; 8, Handling; 9, Water Retrieve. These categories then have a weighing factor according to their importance: (1) × 8, (2) × 6, (3) × 6, (4) × 12, (5) × 4, (6) × 6, (7) × 4, (8) × 4, (9) × 6 in a Novice Stake and × 8 in an Open or All-aged Stake. As there must be two Judges the marks are totalled for each category and then divided by two—the final totals are added up and it is possible to grade the dogs according to their percentage of the total possible marks attainable. At the end of each year these results are published in the Club Year Book and it is interesting to note the variations in the Standards of the winning dogs. No first prize is awarded in an open or All-aged Stake unless a minimum percentage has been acquired. In order for a dog to become a Field Trial champion he must have won two open or all-aged stakes, one of which must have been open to all other breeds which come under the Kennel Club Category of Breeds which Hunt, Point and Retrieve.

Many G.S.P.s in this country are used for all these purposes and these are, I think, the happiest in all probability. They take great joy in hunting for long periods at a time when game is scarce or even non-existent, over moorland, stubbles, grass or in woodland. Pointing is natural when scent is available, and retrieving gives satisfaction to both dog and owner. A few are used primarily as non-slip retrievers. This, to my mind, is better than no work, but a very poor substitute for the real thing for which they were bred. If, in fact, someone asks me for a puppy for this type of shooting I try to persuade him that a retriever would be a far more suitable dog. Wildfowlers use them and get great satisfaction and pleasure, but there again, if a wildfowler intends to use his G.S.P. only for this sport I would suggest that he has a dog with a thicker coat. A G.S.P. will willingly go and sit for hours in water but somehow I cannot approve of this too frequently for a dog with no undercoat.

Falconers use them, and in fact in the last few years it has become a very popular dog in England for falconry. It amazes me how a dog trained by me, with no experience of work with birds-of-prey, can within a few weeks settle in with a falconer and accept his birds and the work expected. I suppose it is possible that the dog's original instincts return, as many centuries ago hunting game with a bird and dog was one of the methods used for acquiring food.

A number of shooting men use these dogs for deer shooting or stalking and they have proved to be extremely useful both in finding a dead animal and also tracking a wounded one which, without an efficient dog, would have been impossible. But if a G.S.P. is going to be appreciated to the full, then it is better that he should be asked to perform all the tasks for which so much thought and care has been taken in his breeding.

Training these dogs is a specialist job and best done by the owner. So much has to be learnt by both dog and owner that this is undoubtedly the right answer. Professional Labrador and Spaniel Trainers are unwilling to accept pointer/retrievers for training. As a pointer/retriever trainer myself I can well understand this. Every dog must be done individually, and both aspects of the training must be taken separately. It is a long slow training, every step being perfected before progressing to the next. Providing the pup is bred correctly, and by that I mean that not only the parents but also grand and great grandparents should be proven working dogs, then the instincts to hunt, point, retrieve and swim should be in-born and all the trainer must do is to control these instincts. But nothing in this world is certain and even the most immaculately bred dog can prove gun-shy or hardmouthed, if not by inheritance then by man power.

I like to see pups of seven to eight weeks carrying things round, even if it is your best hat, a tea-towel or a nestling. At this age I expect them to point anything that is suspect, either by sight or scent—frogs, butterflies, a thrush on the lawn, or pigeon in the park, sometimes even each other. Directly the weather is warm enough they should be introduced to water, first shallow and then deeper, preferably with the example of an older dog. Gunfire should be heard as early as possible, but at a distance and frequently so that no notice is taken of it. Thus by four or five months of age you can tell whether it would appear that your puppy has the potential of a pointer/retriever.

From now on *basic* obedience in all things is the order of the day. Walking to heel, sitting, staying and dropping at a distance can all be perfected so that they become a habit. Simple retrieving can be started in moderation, and introduction to gunfire in conjunction with retrieving. A young dog should be allowed plenty of freedom and exercise where there are no temptations to do wrong. Training should take place on barren ground with no diversions. Once the pup is actively willing to please and attend to your commands then quartering can be started. But it is necessary to choose your ground carefully and ensure that you work into the wind. Many pups will quarter naturally and need only to be turned or called in if going too far ahead. It is vital that you should insist on close work at this stage, nothing is easier than to get a dog to increase his pace and distance, and nothing harder than to get a dog to work close and under control if it has been allowed to go too far and too fast as a youngster.

Probably the most important lesson for this type of gundog is the *drop*. Providing your dog will drop instantly on command then any unforseen incident can be controlled and work started on ground where there is some game. If possible it is advisable that the trainer should be aware of game before the young dog in order to be able to control the reaction. If any type of game holds, then very few pointers will want to move forward once having located the scent, but if the game moves so will the dog and, unless the handler is in control, chasing can occur. At all times and under all circumstances discourage your dog from chasing anything once training has started in earnest. This can be done by forethought, by avoiding the situation or by dropping the dog immediately.

I get annoyed when people ask me how you train a dog to point. As I have said previously a dog should take up a pointing stance naturally and as it grows in age and experience it will discern what needs pointing and what can be ignored. In the beginning encourage your dog to hold his point regardless of the object of the point. This will teach him that you are pleased with his action. At this stage under no circumstances allow him to flush whatever he is pointing. Directly he comes on point approach him from the side, never from the rear, slip on his lead and stroke him quietly. Remain thus for as long as you have patience or as long as nothing moves. After five, ten or even fifteen minutes, if the game has held, hold the dog steadily and push the game up with your foot, as it flushes make the dog drop.

It is important to remember that you are training both a pointer *and* a retriever, therefore that dog will not only have to flush the game but also mark it if it is shot. It must be able to watch the flushed game. This exercise should be repeated as often as possible and for as long as possible so that it becomes a habit for the dog to find game, point, and be 100 per cent steady to flush. Do not introduce a shot until you have achieved this. A shot can be used when doing the retrieving exercises which at this stage should be kept entirely separate from hunting and pointing. Do not associate pointing and retrieving with a young dog, if you do you will get both an unsteady pointer and an unsteady retriever.

Having completed the dog's training thus far it is now possible to take him out with the gun under a handler's control. I say this because it is not easy to concentrate both on your dog and shooting the bird. You will have your dog for ten or eleven season's shooting so it is worth sacrificing the first year of its working life to correct training and ensuring that you have a really good dog for the rest of his time on this earth.

During his first season there are two pieces of worthwhile advice to follow, firstly do not let him retrieve wounded game if at all possible. It could make him hard-mouthed and it certainly will encourage him to chase live game if you encourage him to follow a seen runner. He is too young to differentiate at this stage between the blood scent of a wounded bird and the sight of a bird running ahead of him. Secondly, do not allow him to retrieve game which he has pointed and seen shot and fall in front of him. By this I do not mean that he cannot retrieve that bird, but divorce the point from the retrieve by marking the fall well yourself, then walking away with the dog to heel and waiting some time before telling him to go and fetch it. This will not only discourage him from running into flush or shot but will also improve his memory.

By his second season the dog should be ready to fulfil his role as a pointer/retriever and to have gained sufficient experience in both pointing and retrieving to begin to be really useful. To own one of these dogs which has been trained sensibly and behaves both intelligently and well is the greatest pleasure that any dog-owning shooting man can wish. When a dog comes on point the thrill never wanes, and as both dog and owner grow older and more conversant with each other, it is possible for an extraordinary form of communication to exist.

My original bitch Heidi was an invaluable example of this. We would enter a field and I would cast her off, keeping one or two young dogs sitting at the gateway. Heidi, with head held high, would cover the ground, adjusting her pace to the wind and scenting conditions. Immediately any whiff of scent became available she would hesitate and consider. If game was in front of her she would stiffen instantly and rigidly and, with a flick of her eye to ensure I was watching, by the angle of her head I could tell how far ahead the game was, perhaps 10 feet perhaps 80 yards; by her tail I could tell whether it was still or moving, a slight flicker of her tail meant movement, if it was fur more often than not she would raise a hind leg—but this was not infallible—if it was partridge not a muscle would move, but if pheasant she would allow herself to check with her eyes whether I was approaching. For feather a foreleg would always be raised.

Once she was sure that I was in communication with her then she would decide on whether movement was necessary; she would remain staunch if she considered the game was within gunshot, if not she would begin to stalk forward slowly and firmly, looking back to see whether I wished her to work this point out. If I decided that there might be other game between us and the scent she had located I would give her a low hiss and she would then continue to quarter the ground, always acknowledging the far scent as she passed it. She would always hold her point long enough for me to bring up the youngsters and encourage them to back her, or, if necessary, to wind the game themselves. Then, when the scene was set, she would flush and mark until whatever it happened to be was out of sight; only then would she turn round to her class as if to ensure their complete attention. She knew more about game and shooting than many a gun she served.

An American friend who had experience of many G.S.P.s, both in the States and Germany, borrowed Heidi for a month's grouse shooting at Hunthill and later wrote saying 'to be perfectly truthful with you, Heidi was the greatest prospect German that I have ever worked. I believe Heidi's strain should be carried along as she showed as much bird sense as any dog I have had the pleasure of working with. Too many G.S.P.s have now become inbred, nervous and soft'.

Heidi was not the ultimate in steadiness or obedience, but that was my fault; nevertheless she was always in great demand on any shoot I went to because of the unfailing sport she provided for the guns. She would find game where no game should ever be, and would retrieve game that many a gun swore he had never touched until I showed him a pellet in the wing or leg. Just once or twice, or maybe more, she did bring back a bird unwounded but that was only because she reckoned if it was stupid enough to allow itself to be caught it must be mad or ill. She had a great sense of humour and an enormous love of chocolates. Even in her very old age, when she was both deaf and blind, it was not possible to extract a chocolate from its wrapping, no matter how furtively you did it, without her hearing.

Her greatest joke, which never ceased to amuse her and irritate me, was when I would expect her to cover a barren field and produce game where no self-respecting game should be. She would obediently quarter it to perfection getting further and further away from me until she reckoned she would cause the most annoyance. Suddenly she would swing into a back point with all the signs of excitement that scent could give. I would inwardly curse, having given up hope of any sport, but start approaching her hurriedly though quietly and attentively. When she reckoned I had suffered enough she would give a single twitch of her tail and dive into the grass, coming up with the tail of a field mouse or the leg of a frog sticking out of the side of her mouth, and an enormous twinkle in her eye. She never failed to catch me out on this game, as she knew well enough that I lived by the old adage of 'trust your dog'—certainly where she was concerned.

Each dog differs and has to be dealt with differently. I've had a ten-month-old dog puppy win a field trial and a two-year-old bitch who is only now beginning to accept training, a few years ago I had a black bitch ready to be shot over when she was a year old, and a dog in for training who, at ten months, had to go home again as he had the mentality of a pup of four months. Of course some of this is in the upbringing of the dog, but most of it stems from the individual temperament and it is of the utmost importance to adjust your pace of training to this.

Make haste as slowly as you like, don't bore your pupil or yourself and do not physically chastise a German pointer unless you have really good cause to do so. To my mind this can only be excused where danger to the dog or a person is concerned and all other methods of correction have been tried and failed. Normally an angry word, gesture or shake is quite sufficient to enable the dog to understand that his action has displeased you, and as willing obedience is your code word, then the lesson will have gone home, providing always that he has understood what it was that he did wrong. In order to ensure this, if you re-make the situation immediately after, giving the dog the opportunity to make the same mistake, and he does not—you have the right answer, but if he repeats the mistake then the chances are that he has not completely understood.

Dogs are not normally deliberately disobedient, they will try to avoid doing what they are told if inconvenient, but that is different. Only when I see a dog looking at me, interpreting my command correctly, and then deliberately disobeying, am I sure that correction is necessary and will be beneficial. So often a dog's disobedience is lack of understanding and in a breed where the nose work is of such importance I am convinced that, just as when a human is concentrating on some interesting book, sounds, noises and sometimes even contact do not have any effect, similarly when a dog's nose is in full use his ears are blanked off. I know this may seem a silly excuse to many, but I am firmly convinced that it is a fact, primarily because it has often been the pattern of behaviour in many of my dogs who would rather lie down and die than cause me displeasure.

Mating, whelping, rearing, weaning, feeding and caring for German Pointers is, to me, the normal natural way that any sensible person would look after a dog. Mating and whelping should be natural with no interference whatsoever. If it is needed then either the mating should not take place or the pups that cause difficulty are better dead anyway. Only once was one of my bitches force mated, without my consent, and the result was dead puppies. At times there have been weaklings in a litter and I have never made any attempt to hand rear them, on the principal that if they are weak then they will remain so all their lives in one way or another, and also that if their fight to survive succeeds then the chances are that they will be among the best of the litter. A bitch does not discard a puppy lightly, and when she does there is usually some basic fault. Litters vary from five to twelve, an average I would say would be eight pups, and a maiden bitch is quite capable of rearing that number, but not more than eight unless you are prepared to start weaning at three weeks in order that she will not get run down.

But the most important consideration when deciding on the number of pups to rear is the availability of suitable buyers. This is not a breed to be sold regardless to anyone who thinks he would like a G.S.P. because their appearance appeals to him. Ten years ago I seldom heard of an unwanted G.S.P., nowadays it is happening more frequently. As a young dog grows and matures in size and intelligence so do the necessary understanding, time and training requirements increase. This must be carefully and fully explained to any potential owner and, unless they are prepared to accept this responsibility, they should not buy such a dog.

Finding the right homes for second-hand dogs is incredibly difficult, particularly when a dog has been spoilt and pampered as a puppy and then neglected and discarded as he becomes a nuisance, entirely due to lack of attention and appreciation of their ability and intelligence. In so many cases I do not blame the owners but the breeders who have sold the pups regardless of a careful scrutiny of the buyer's intentions. It is far wiser to rear a few puppies which can be assured of happy, suitable homes, than a dozen that will perhaps bring in the money but by a year old be unwanted and untrained. The first year of a dog's life is the most formative and much cruelty and unhappiness could be avoided if breeders were more selective.

A bitch, after having reared a litter, should look as fit or fitter than she did before. I like them to be fully weaned by six weeks, but allow the bitch to sleep and play with them, providing they are not too rough, for another three weeks, as they learn much from her at this age, in particular reaction to gunshot.

If they are all together, playing in the open, I will fire a shot 60/70 yards away; the bitch will show excitement and interest and it is amazing to see how the pups will copy her. She can also influence them in a love of retrieving and interest in, not fear of, dead game at this age.

As with the rearing of any young creatures economy does not pay. Only the best is good enough; they need plenty of good food and in this way you will produce pups with bone and substance that you can be proud of. No matter how good the meat or how wholesome the biscuits, extra vitamins and minerals are still necessary to this breed and should be continued until they are fully mature—that is till between 18 months and two years. During the weaning period they should be fed five times a day, but at eight weeks this can be cut down to four, then three meals at 12 weeks and from six months to a year they should be fed twice a day. After that it is usually practicable to give only one meal, but I often find that an energetic youngster does better on two small meals than one big one, probably until it is two years old.

Round worms must be accepted but rather than wait for an infestation it is better to worm the dogs regularly three times a year. As with all other breeds G.S.P.s do moult, despite the fact that their coats are short and need the minimum of attention. Undoubtedly the coat grows coarser and thicker if the dogs sleep in unheated kennels and the natural twice yearly moult occurs. But in this modern world of centrally heated houses nature is befuddled and most dogs living in the house will tend to moult continuously all the year round. After all, it is only nature trying to produce the correct clothing to suit the temperature. Despite this I am a strong advocate of this breed of dog being kept in the house. So much basic training and rapport with the owner can be achieved, both during the day and in the evenings, and so much more enjoyment can be had by both owner and dog if they are together. That the G.S.P. is a gundog is not to be doubted but that it is also a family friend, guard, clown and companion is a facet which gives both owner and dog the greatest enjoyment and pleasure.

To date the breed is virtually unspoilt in this country; many show dogs are worked and many working dogs are to be seen at shows. This is proved by the fact that they are one of the few breeds which have Field Trial Classes at Shows and these classes are well supported. How long this situation will last I would not care to say, for as the breed gains popularity, so it will decline and I fear, inevitably, we will have the pretty elegant show specimen and the more substantial, heavier-boned dog with the stamina capable of doing five hard days' work a week. It is the duty of those owning and breeding working dogs to see that they conform as nearly as possible to their interpretation of the standard of the breed regardless of the preferences and peculiarities of the show judges. It would be of considerable help if we could emulate the Germans by having a test of conformation incorporated into our field trials to ensure that the top winning dogs in the field are truly representative of the breed and worthy to be bred from by selective breeders.

18 Weimaraner

This outstandingly handsome grey dog, known in the States as the Grey Ghost, has a very special temperament and character and its history is complicated and devious. Two main theories stand out, one that any documents pertaining to the breeding of these dogs have been lost during the wars, the other, that the details of the breeding were kept a strict secret so that it should be kept pure and indulged in only by the selected few. It is possible that both theories are true; that the secrets of the breeding were kept privately and that these were either mislaid or destroyed during the wars.

Delving through the details given by experts on the breed it is difficult to assess whether the claim that this is the only shorthaired pointer retriever that has not been cross-bred and has remained virtually the same as seen in pictures of the 17th century where shorthaired grey dogs were depicted in hunting scenes, or whether, like the German Shorthaired Pointer, it has been evolved from hounds and pointers in order to produce the necessary all-round working ability required by its masters. The theories of the history are most interesting as they involve the most distinctive characteristic of this breed, that is the colour, described as varying from silver grey to shades of mouse or roe grey. The experts who claim that the breed does not carry the blood of pointers, in particular German pointers, hold the theory that this colour was produced by mating two black dogs and then when the grey colour appeared, which it is bound to do when two blacks are mated, these greys were mated and so the colour was tied. The black dogs were the St Hubertus Brachen, a black hound bred by Benedictine monks in the Ardennes as long ago as the 9th century. In support of this idea the tan markings on the eybrows and legs which appear from time to time in one or two puppies in a litter would indicate a throw-back to the hound. Alternatively, a silver grey can also be produced by the fading of brown-coloured dogs as the result of a deficiency of vitamin D, and if the conformation of the Weimaraner is studied carefully although there are very definite differences both in type of head and body, equally well the general outline resembles the G.S.P. fairly closely. The Great Dane is also mentioned as a possible ancestor.

Nevertheless, we do know that as the methods of hunting changed with the development of weapons so did the demands of the hunters for the type of working dog required. With the advent of firearms came the need for a bird dog which would range ahead, point and hold game until the guns could approach within gunshot. Thus, from the pure-bred tracking and trailing hound of a heavy build came the introduction of the pointer with its greater speed and lighter bone.

In 1775 Grand Duke Karl August, Duke of Saxe-Weimar-Eisenbach came to power (he died in 1828). He was a very great admirer of the grey hunting dogs and of their abilities in the field, and entrusted their breeding development to his nobles under the strictest secrecy in order to maintain pure and selective breeding that held no place for profit or commercialism. The name of Weimar comes from the town, the capital of the State of Thuringia, a city of both beauty and culture. The nobility had access to some of the best stocked shooting estates in Europe, so wished to breed and train a type of dog which would please them aesthetically and also possess the intelligence required for their field work. Previously the breed is said to have been used in packs for hunting big game such as wolves, wild cat, deer, mountain lion and bear. Now it was required for the small game that the huntsmen enjoyed shooting for sport. Nose, speed, courage and stamina were necessary to enable the breed to find, hold and retrieve the game from land and water. Obedience was greatly stressed and in books on the breed in those days it is interesting to read that the sensitivity of their characters was noted and care taken not to handle them roughly.

In 1897 the Weimaraner Club of Germany was formed. This was an exclusive club whose aims were the improvement of the breed. There were firm restrictions both on membership and ownership. Breeding was strictly controlled. The emphasis was entirely on the field work of the dog. In 1913 the first Weimaraner was imported into Austria and great care and interest has been taken with the breed in that country ever since. In 1929 two Weimaraners were allowed to be exported to America and after the last war, in the early '50s, they came to Britain.

The publicity given to the breed was unfortunate, for it was claimed to be the wonder dog of the shooting world—it may have been so—but this was not backed up by proof in this country. The conservative Englishman was not pleased to have it suggested that a Croat was capable of superiority in the field over the long established and well loved retrievers and spaniels. The genuine admirers and supporters of the Weimaraner, who knew and appreciated its ability, were over-enthusiastic and instead of claiming that, in fact, it was a truly useful gundog in its own right for those shooting men who required this type of gundog, the inference was that it could outshine the other established gundog breeds in the country and this was taken as an insult. Hence antagonism to the breed became vociferous in the press and much harm was done at this stage in England. A major criticism against the breed was that they were cat killers, but the Germans encouraged this as cats, either wild or domestic, do as much if not more harm to game than many other predators. But the cat-loving British public did not approve of this. However, it was hardly fair to lay this crime at the feet of the Weimaraner breed alone. There are cat-killers in every breed of dog from the smallest to the largest. If inherent this trait can be controlled and if developed it can be stopped. I have always kept a family cat, not because I like cats, but because I do not like cat-chasing dogs. A ginger Tom is the best deterrent for anti-pussy puppies and although they may never grow to like cats, it will instill a healthy respect for them, respect they should have for any domestic animal.

Happiness was restored to an old-age pensioner neighbour of mine entirely due to a Weimaraner. She came to me in tears because her cat, which had been ill, had not come home for three days and she had given up hope of finding him. I said I would take a turn up the lane and round the hedgerows in case he was lying up dying. I took Ric with me, who was boss dog at that time, and we set out in the twilight. We had not gone far from her cottage when he went on point near a stump of a dead tree in thick nettles. Peering in I saw the cat curled up and very miserable. We took him home, called the vet and within a week both cat and neighbour were fully restored to happiness and health again.

Just as an Englishman's home is his castle so aspersions cast on his dog can be one of the greatest insults. The Weimaraner has had to live this

72. Weimaraner bitch belonging to the author.
Peter Moxon

down over the last 25 years. Despite it many dogs have been bred and quite a number have been owned and enjoyed by shooting men, both on the moors in the north and in the pheasant country further south.

A bitch called Misty has given a lifetime of good service to her master. I have often been out on shoots with them and never ceased to be impressed by this Weimaraner. Her hunting, pointing and retrieving was excellent and she was always in great demand, as was her owner who had that wonderful gift of remembering funny stories. The pair of them not only produced, shot and retrieved the game but also entertained us all. One of Misty's party tricks was to 'sing' on command. At a dull moment on the trailer or when an angel passed overhead in the pub, Harry would say 'sing Misty' and up would go her head and a wonderful howl of the Baskervilles emerge to the consternation of any strangers. Her other party trick, which I strongly depreciated at times, was an ability to jump on command. She would go backwards and forwards over a five-barred gate without hesitation, or any obstacle for that matter, when requested. But I must add that her owner always had in his pocket a needle and thread as there were times when, in her haste whilst working, an unseen strand of barbed wire would rip her skin and he could be seen sewing up the damage on the spot.

But alas, I fear many have been bred purely as pets or as show dogs. The glamour of his colour and the beauty of his conformation have had a great hand in the downfall of the Weimaraner as a working gundog in this country and this is also true in the States.

The Standard of the breed requires a medium-sized, silver grey dog with light eyes. His temperament should be fearless and friendly but protective and obedient. He should give an overall impression of power, stamina and alertness. Movement should be free, indicating co-ordination and balance. The head is described as aristocratic, moderately long, a prominent occipital bone, high set, long ears and a moderate stop. The jaws should be powerful and the neck clean, long and muscular. The teeth should meet in a scissor bite, nose should be grey and the lips and gums flesh colour. The body is well muscled compact and square. The feet firm and cat like, dewclaws to be removed. The tail is docked by approximately just over half its total length. The coat is short, smooth and sleek giving a metallic sheen. A small white mark is permissible on the chest. Height from withers to ground for the dogs is 24 to 27 inches and for the bitches 22 to 25 inches.

No mention in the Standard is made of the long-haired Weimaraner. It would appear that originally the long hair was a mutation within the shorthair, one or two puppies in a pure bred shorthair litter having a longer silky coat. At first these were rejected but at the end of the 19th century they were being bred purposefully for their excellent work in woodland. In 1935 the German and Austrian Clubs approved a new Standard for the breed which included longhaired Weimaraners provided that their origin could be traced back for four generations. There is no distinction between the dogs other than length of coat. The tails are undocked. Because of their scarcity they disappeared completely during the last war but have reappeared since and in 1950 one was exported to America and now, in the last year or two, they have made their appearance in this country.

The character and working ability of the Weimaraner has greatly influenced its development. This may seem a very obvious statement, but it is so noticeable in this breed, that explanation is necessary. The character of the breed is something that enthusiasts never tire of extolling. The head is described as aristocratic, as also is the character. These dogs are loyal, affectionate and friendly to their family, but they are aloof to visitors and aggressive to strangers. This surely is the result of their environment when raised by the nobles of Weimar. They were not kennel dogs, but part of the family in the castle and as such had the duty of friend and guardian to perform as well as their work in the field. Possessiveness is a strong trait in the breed and has to be handled carefully. I would say they consider themselves to be a breed apart; they will fight other dogs if their superiority is questioned, but are quite happy to live amicably on the understanding that they are boss. I have owned and trained many and have noticed this attitude in all of them. They like their own kind and will play happily together but remain aloof from other breeds. When working with other dogs they need no training to work independently. They are not demonstratively affectionate as some other pointer retriever breeds can be, but, nevertheless, give a loving loyalty that is not easy to describe.

Some years ago I exported a puppy to an American family living in Turkey. They required it as a housedog and also for shooting. I had good reports about the dog, particularly concerning his working ability. The last letter I received told me that because of this dog's warning not only their house but also their young daughter had been saved from a fire which had started in the basement. The dog wakened them and then jumped through their daughter's window on the ground floor to enable the child to escape.

I only know that, although I do not rate them as being the ideal in the pointer/retriever world, I would not be without one in the house as a friend and companion. As far as their natural ability to work is concerned I am a strong supporter of the theory, previously mentioned, that basically they stem purely from the St Hubertus hound. Their tracking and retrieving instincts are very strong indeed. As puppies it is very noticeable if you have a mixed collection of Weimars and G.S.P.s running on together, that the G.S.P. pups will hunt with head held high searching for air scent, whereas the Weimar pups will run noses to the ground following any ground scent with perseverance and speed. As they grow on it is necessary to *train* the Weimar to get his head up and the G.S.P. to get his head

73. Friedrich Von Weimar, first Weimaraner to win a G.S.P. Field Trial.

74. 'Can't be too careful'.

down. The result of this is that although it is possible for the Weimaraner to be seen working at speed, covering his ground well and using his nose to the utmost, when scent is available for him to come on point it is always at close range. Over many years I have never owned, trained or judged a Weimar who has acknowledged a point of 80 to 100 yards such as one would expect from a German or English pointer. This may not be considered necessarily a disadvantage; they are certainly natural pointers, and are staunch at point.

They are inclined to work more slowly and more closely than the faster, wider ranging pointers. But undoubtedly those bred from working stock have excellent noses and many a time when game has proved to be totally lost to other dogs, a Weimaraner has produced it. I have known a number of them to be hard-mouthed but who is to say that this is not bad training or bad breeding from acknowledged hard-mouthed dogs? I have known just as many with the softest mouths possible. They are naturally good, strong, water dogs if introduced to water early. In fact, in Germany this was one of their greatest assets. They jump naturally and well and their tracking ability combined with stamina and perseverance make them formidable hunters for any wounded game. We all have our likes and dislikes, but I have yet to meet a shooting man who has owned a well-bred, well-trained Weimar and who wished to shoot over any other breed of dog.

Their basic training is much the same as that for the German short-haired pointer, early obedience being of great importance, and fairness and sensible kindness being essential. In order to encourage them to work with their heads held higher than appears to be natural, it is wise, when teaching them to quarter, that this should be done on ground where either grass or stubble is at least nine to twelve inches high, and ground scent is not available if possible. In other words, do not work them where hares or rabbits are likely to be or have been. If they start foot scenting drop them immediately and cast them off again on a different line. As with the G.S.P. do not in the early stages combine pointing and retrieving but start waterwork early.

One interesting point that comes to mind is that among my own or other people's dogs which I have had to train I have never had a gunshy Weimaraner, nor have I had any form of gun nervousness. I have also had a number of tree-climbing Weimars! Some were more adept than others, and although dogs do teach each other none of my other breeds have learnt to do this. An old dog I have now will still climb at least 20 feet up an oak tree which overhangs a pond. In the wet weather more than once he has slipped from his favourite branch into the pond much to his irritation and my concern. This convinces me as much as anything how necessary it is that no dog should wear any form of collar whilst working. I am constantly surprised at the pictures of continental trials in which the dogs appear to be wearing choke chains.

Once again the care, feeding and breeding of Weimaraners should be as natural as possible. They need plenty of good food, plenty of free exercise, and a natural mating. The urge to merge, as in any other breed, is normal but a randy gundog is a bore. The Weimaraner Stud dogs I have owned can never be described thus. If a bitch comes for mating and is ready then nature takes over quietly and successfully. I well remember one occasion when I was getting ready to go shooting, clothes and dogs were all aboard, then a car drove up. I had completely forgotten that a bitch was coming for mating that day and I felt appallingly guilty. I apologised to the visitors and called Ranger out of the Land-Rover. He obligingly came and said 'hello' to the girl in question and then 'hello' again. For absolutely no reason, that I now can remember, at that moment I shut the door of the Rover. In a flash Ranger was in the back, still drooling, but with not the slightest intention of allowing women to supplant a day's sport. I took trouble to explain to the owners that a marriage was unlikely until later that day, and if they cared to wait we would be back about 5 o'clock.

We had a very good day; Ranger found two woodcock that other dogs had missed, and retrieved more than his share of the 60 or more duck that fell in the water and, as so often happens, as we were finally loading up to go home, he went on point right in the hedge by the cars. I hadn't yet missed him and it was one of the guns who called me over. Pheasants are curious birds, often they appear to have no fear whatsoever of the human voice, machinery or even dogs.

This situation has happened time and time again. Before a drive I can be standing talking to a couple of guns and laughing away with the dogs poddling around 'officially' at heel, when one will suddenly freeze within a few yards of us. In these conditions I insist that they come off point as it is unnecessary at that moment to flush that bird. People think this is bad training for a pointer, but I have found it to be essential in the work I expect from my dogs. In the spring if a hen sitting on her eggs is pointed, I certainly do not want a flush. On a drive waiting for the beaters to come, I do not want a flush, nor is it necessary at the end of the day, when all the guns have been put in the cars, and the shooters themselves have had enough, for that bird to be disturbed.

A pointer/retriever must have two commands when on point—either 'get in' i.e. to flush, or 'leave and come'. I have never yet spoilt a dog by insisting on this.

Anyway we did flush and shoot that last bird as a visiting gun was sceptical of its presence and I fear I could not resist proving my point! We got back home to find Ranger's 'intended' waiting. He jumped out happily, still full of energy and enticed her off into the field by the drive—Half an hour later he was enjoying a rather larger meal than usual, the bitch was back in her car, and her owners were sharing a long awaited and much appreciated drink with us by the fire. Sixty-three days later a fine litter of baby Rangers was produced. What a perfect day for a gun-dog—sport, sex and steak—and in that order. But seriously I have never had trouble with my stud dogs. They do not bother bitches at all on the shoot, nor do they drive themselves up the wall when one of our own bitches is in season. Round about the 10th to 12th day I admit I have to suffer morning and evening prayers in almost Arabic style—but otherwise if their services are required—they are given sensibly and with obvious enjoyment—and if they are not then little or no umbrage is taken. Peace

75. Weimaraners mixing happily with other breeds.

and quiet are the only necessities when whelping. They are easy whelpers and the bitches make naturally good mothers. A fat, sturdy litter of Weimaraner pups is a glorious sight. Their pale grey coats accentuate their periwinkle blue eyes. It is no wonder that so many are sold to totally unsuitable homes.

Proof that there is a split in the breed in this country, between working and Show dogs, lies in the proportion of Show Champions to full champions. This has undoubtedly led to some deterioration of their working ability in the field. But a number of them have excelled in obedience and working trials and some have been used successfully by the police. It is of interest to note that Weimars and G.S.P.s were used in America to locate parts of explosives devices necessary for examination after the explosion had taken place. The parts were coated with whale oil, in order to enable the dogs to point them. Despite the split in the breed which has affected the working ability, it is noticeable that the conformation remains unspoilt.

The Weimaraners I have exported to Scandinavia have been solely for shooting purposes and have been greatly appreciated. These dogs have all been very much part of the families they joined. I had an excellent photograph sent to me of Abdullah who went to Copenhagen; he was lying in the sun on a deck chair with an English cup of tea on one side of him and a girlie magazine the other. Later in the winter they sent me another photo of a shooting party in the snow with six Weimars. They use them not only as pointer/retrievers but also as pack dogs. The party drives as far up into the mountains as possible and then loads the gear, each dog carrying about 40 pounds in panniers over his back. On arrival at the shooting lodge they are unloaded and revert to being gundogs again. In fact the breed has done well recently in the sleigh races for, being short-coated, the snow does not pack up on their feet to impede their speed.

There are good working dogs earning their living and giving pleasure and satisfaction to their owners. These will no doubt continue and be bred sensibly so that in future years the Weimaraner will take its place and succeed in pointer/retriever trials. I won a G.S.P. Field Trial with a Weimaraner some years ago—although my best friends remarked at the time that the standard must have been very low! Since then two bitches have had first awards so there is still time and hope for a gundog whose breeding was developed by perfectionists who needed and produced an all-purpose gundog. The Weimaraner still has all the potentials of such a dog if bred and trained correctly. It is well worthy of regard and commendation in the shooting field by discerning owners who desire something different and rather special.

19 Hungarian Vizsla

The Vizsla or the Golden Pointer, as it is sometimes called, like the Weimaranar has the distinction of a specific colour which has considerable bearing on the evolution and history of the breed. The Vizsla's original ancestor was most probably a cross between the native Pannonian hound, an ancient Hungarian foxhound, and a yellow bird dog that was used for hunting birds with either weapons or falcons by the tribes which emigrated from Asia, crossed the Carpathians and eventually settled in Hungary about A.D. 1,000. The Magyars or Hungarians named the breed after a small village called Vizsla in the Danube valley.

By the end of the 19th century, after introducing the blood of a variety of other hunting dogs such as the bloodhound and different pointing breeds, the 'golden pointer', as we know it today, appeared. The hunting requirements controlled its development. The plains of Hungary consisted of vast areas of cornland which held a large variety of game birds and a breed of hare which was particularly large. Thus the shooting dog for this particular type of country had to be fast, active, persevering and capable of holding the game. It had to withstand the very hot summers and bitterly cold winters. Not only was it expected to hunt and point the game but also retrieve it from land and water. The varying shades of its golden coat enabled it to blend in with the terrain and by cautious and intelligent hunting it would lead the shooters to within range of the game.

An interesting method was used for producing the giant hares for the guns. The Vizsla would flush the hare and then chase it until finally it circled round towards the line of guns. When the hare was within gunshot the dog was trained to drop in order to allow the guns to shoot.

After the First World War the Hungarian Empire was diminished and much of the territory was divided between Rumania, Austria, Czechoslovakia and Yugoslavia. The breed became very scarce, only preserved by its most ardent admirers. During the interval between the world wars selective breeding took place by the owners of the large estates still left in Hungary. This was entirely without commercialism, which ensured the rearing of first-class hunting dogs of the standard required for the work expected of them. They were used in packs to hunt and hold at bay wild boar and wolves; they were used by falconers; they had to find the giant hares and drive them towards the guns; they worked deer in the forests, and pheasant and woodcock in woodlands. They had to find and point quail in the marshes, flush duck and geese from the river banks and then retrieve whatever game fell to the guns.

Since there was no financial gain selective breeding could be indulged and commercialism kept at bay. Bench shows were strongly discouraged as this, it was felt, would without doubt lead to the deterioration of the breed. This is why the breed was virtually unknown outside the Hungarian Empire in pre-war days. However the breed was again sadly affected by the last war and during the Russian occupation was almost wiped out. Fortunately it was officially recognised by the Fédération Cynologique Internationale in 1935, and in 1938 was first exported to America. After the war more found their way to the States and in 1952 two were imported into England via Denmark. In 1958 they were recognised by the Canadian Kennel Club. Our basic English stock has been bred from the original pair which came from Hungary, an American bitch whose parents were bred in Czechoslovakia, two more dogs imported from Hungary and a dog who came from France but whose sire was bred in Austria. Since then, dogs have been imported from the States, Hungary and Austria. But it is interesting to note, when looking at an extended pedigree of all these dogs, which numerically are not very many, that they all share common ancestors from either Hungary or Czechoslovakia. From photographs of the breed in these two countries it is noticeable how, although the type remains the same, the conformation differs considerably. The Hungarian dogs appear to be very much lighter in bone and smaller in size whereas the Czechoslovakian dogs are heavier and stronger in substance. This difference was quite noticeable in our basic English stock and it would seem today that by judicial and sound breeding we have produced a type that has both substance and quality which conforms to the accepted Standard of the breed and compares very favourably with the best that has recently been imported from Hungary.

There is much in common in the Standard of the Vizsla with both the German shorthaired pointer and the Weimaraner. It is a shorthaired dog of medium size, with power, stamina, intelligence and a tireless determination for work. It ranges in colour from a sandy yellow described by the Hungarians as 'crust of loaf of bread' to a deep russet gold, but must not resemble the dark red of a bloodhound. In conformation Vizslas are compact; the length of body should be the same as the height from the withers to the ground, but they are lighter boned and of a more lithe and graceful build than either of the other breeds. The tail is docked by only one third, and one presumes the reason is in order to avoid damage. The ears are set low and fairly long. Eyes, nose, lips and nails should tone in with the shade of the coat. There should be no shading or white marks.
 Dogs should stand from 57 to 64 centimetres, and bitches from 53 to 60

centimetres. Weight should be between 22 to 30 kilogrammes.

The greatest difference between the breeds of pointer/retriever, lies in the temperament. The Vizsla's temperament is an anomoly—utterly contradictory. Happiness, confidence and a deep and sincere wish to please are all evident and genuine. Nevertheless, he is also sensitive and contrary—soft and very strong willed. This creates a strange and difficult mixture for training, but for those who understand them it is utterly fascinating. Where facing cover or strong running water is concerned they are fearless, but facing a strange polythene bag they can be as timid as any field mouse. If scent is strong they can be carried on the wings of the wind regardless of the most fearsome cursing, yet if handled or corrected roughly can become a spineless mass of cowering jelly. The Vizsla is neither a fighter nor a biter but he will not accept domination from other dogs or from strangers. A curious trait that I have noticed is that if forced to retaliate to an aggressive dog it will not mince matters but go straight for the throat. This also applies when their guarding instinct is questioned. A bitch I had, who was imported from Hungary, was happy and friendly to all comers except in the car, when she turned into what appeared to be something so vicious that no one in their right mind would approach her, or at night if a stranger came into the house without prior greeting from me. She would slide out of her chair and face them squarely, hackles raised and lips drawn back. Only once did she attack and then went straight for the throat. This compares oddly with the Weimaraner who is known to have an aggressive tendency and strong guarding instincts but when mine have been involved in a fight they always attacked from the rear and come off worse.

When, as quite often happened with a favourite Weimar dog of mine, who was known and feared in the area as a biter, he attacked anyone they never received more than a rather hard pinch in the bottom—a large bruised area but with hardly a scratch on the skin.

Generally speaking, Vizslas can be relied upon to make useful houseguards despite their gentle temperament. Their need for physical affection is strong and more ground can be covered in training by praise than any other method. Their clowning ability causes me constant amusement—they will play games of hide and seek or throw things in the air and catch them, carry anything from a bucket to a handkerchief at full gallop, lie in the best armchair with head hidden down the side ostrich like, hoping not to be noticed, dig furiously for a non-existent mole and then suddenly leap in the air and catch a butterfly.

Somewhere way back in their ancestry there must have been the blood of an entomologist as they are incredible moth and butterfly hunters and as pups will persevere for hours hunting them. A dear dog of mine called Kos at the age of four was leading an expedition after a Red Admiral—he gave chase then a leap with open mouth and the butterfly vanished. He turned to me in consternation, opened his mouth in surprise and out flew

76. A good mouthful well-held by a Viszla.

77. A Viszla marking the fall.

the butterfly to be chased again by the others!

To sum up the general characteristics of this breed one can do little better than to quote the English translation of the Standard as issued by the Hungarian Kennel Club: 'It is intelligent, docile, quiet, but a type of vivacious temper. It has an excellent good nose. It is quiet as it stops the game at bay. It works readily even on moist ground. It is affectionate. It bears very well any guiding. It is obedient but susceptible of bad treatment. It is a manysides hunting dog'.

Training should start early with basic obedience and a firm healthy respect by the dog for his handler and by the handler for the intelligence and willingness of the dog to learn. The dog must accept the superior knowledge and control at all times of his handler. Self-hunting is natural if allowed in any of these breeds but if controlled when young will produce an excellent worker willing to hunt for his handler and not for himself. Pointing is a strong natural instinct, to be fostered and encouraged. Retrieving comes naturally as well and is enjoyed to the utmost by Vizslas. I can truthfully say that I have never owned a hard-mouthed Vizsla, nor have I had one that does not revel in water work even in the coldest conditions. I think that the fact that their coats appear to have a slightly greasy texture has some influence on this. The cold does not seem to affect them, and they will play for hours in the snow despite their apparently thin coats. It surprises me, too, that they show no fear whatsoever of any type of cover and although at the end of a hard day in thorn and bramble their faces and underparts are covered in scratches and rashes this does not deter them one whit the following day. One big advantage they do have over the more hairy types of dogs is that there is no necessity to check for burrs in their coats nor do they need more than a five minute rub down on arrival back at home before they can be cosily curled up in your chair. It is wise, however, to check for thorns as these stick and stay in ear lobes and on their bodies. Jumping holds no fears or problems but it is important in these days of barbed wire to train them to jump cleanly over obstacles to avoid danger.

The same can be said of their working potential as has been said of the G.S.P. but their temperament is gentler and more sensitive—some would say that they are too affectionate. They are a curious mixture, for they are very easily cowed by harsh treatment but have a streak of stubborness that must be overcome if the essential obedience is to be achieved. Kind but very firm handling is needed; dogs, like children, become what we make of them. They say the dog is as good and sometimes better than his trainer so it is up to the owner to ensure that a Vizsla will work willingly for him. My old Vizsla bitch, who was imported from Hungary in 1959, would give any shooting man a most exciting and productive day's sport, either on the moors, marshes or in woodland, and nothing that remotely resembled game or vermin would be missed. She would point and retrieve the lot—but like Frank Sinatra 'She did it her way!' There was no doubt, I fear, who was in charge of the day and like a good gamekeeper she expected you to appreciate that she knew where the game was to be found and you would be well advised to take her advice and see that you kept up with her pace. She taught me the true potential of the Vizsla as an all-purpose gundog, but also the necessity of basic control if one was to go shooting in the company of others. Probably the best retriever I have had or shall ever have was Nefertiti—a little bitch so called because she was no great beauty. She was small and wiry with an inquisitive expression that reminded me of a keen ferret. She was an independent little dog with a restless urge for work. She would be curled up quietly and apparently fast asleep, but should a gun be taken from the rack or the cartridge drawer opened she was by your side. It was useless to try and escape without her company; she had no respect for doors or windows and if all else failed was capable of making such an appalling noise that the most hard-hearted would have relented.

She needed no training as such, she was at heel without a word, in and out of the car or Land-Rover, silent and watching until we arrived at the shoot. Then this little red dormouse would, as if by magic, become the Goddess Diana. No matter how far or fast she ranged she would hold whatever game was there until a gun arrived. She covered open ground like an English pointer and thick brambles as thoroughly as any spaniel. She possessed an in-born knowledge of whether game was dead or wounded. After flushing game for a gun she would pause and then either return with the shot game or continue hunting regardless of the noise of the shot—no, not regardless, because somehow she knew every time whether or not a bird was down despite the fact that it was completely out of her sight.

The most impressive memory of her was on one drive when a bird was shot and dropped dead about 100 yards behind the guns. As all was quiet

I sent her for it. She picked it up but half-way back a cock rose high from the woods above the guns and was pricked. Neffi stood stock still with the first bird in her mouth marking the cock's flight. She then put the first bird down and went straight after the runner which had flown on a long way—I waited—minutes later she returned with the cock, eye bright and tail up, straight to hand. She then returned immediately and collected the dead bird with not a murmur of command from me. That was not field trial behaviour, but practical intelligent shooting dogwork.

At the end of the season I would take her to collect wild hens for the keeper's laying pen. She would point them quickly and then very gently pick them up with not a feather displaced. I shall never know how she knew that during the season she must flush them and at the end must 'peg' the hens for us.

Her world began and ended in the enjoyment of shooting. At the age of four she dropped dead with a hare in her mouth five yards from me. A hind leg had been peppered, she followed and retrieved it alive over plough, up a hill and finally jumping a pig wire fence. She was the runt of the litter and must, unknown to me, have had a weak heart, but in her three seasons' work no dog showed greater potential as a completely natural working Vizsla.

The breeding and rearing of this breed should be done with the utmost consideration for the maintenance and improvement of their working ability. When people buy a dog they are willing to travel considerable distances to find the right one, so why then, when the time comes to breed, are they so unwilling to go further than is absolutely necessary? Is it fair to the breed, or to the bitch herself? Surely, when considering a stud, it isn't the distance and trouble but points which one would like to see improved in the resultant progeny which should decide which dog is to be used. If you have a shy bitch, then avoid using a shy dog; perhaps she is a plodder when searching for game or goes like one possessed; perhaps she is short in the neck, or lacks angulation. These are the points to consider, not distance. We must aim to produce quality at the expense of quantity.

If you wish to raise first class dogs then you must be prepared to take trouble, spend money, and have patience. There is no short cut to success. Study the standard of the breed carefully, remember the working capacity of the dogs instinct-wise, consider type, colour and soundness. Ask your Vet for advice on breeding and get to know your own dog thoroughly so that you can acknowledge the good and bad traits, both dominant and recessive. Remember that foundation stock must have an excess of good points, and the bad ones must only be minor faults. Get a clear picture in your mind of the sort of Vizsla you wish to breed. Never mate two dogs without a definite purpose. Have a reason for choosing your stud and a picture in your mind of the resultant progeny. Pedigrees alone are not sufficient; knowledge of the individual dogs is essential. If this were not true then producing dual champions would be child's play. Generally speaking, line breeding is more likely to be successful than an out-cross (i.e. two unrelated dogs) but remember line breeding never creates anything new—it fixes what is there, so you must be sure that you have all the virtues you require. It can easily establish faults if they are there, and it is well known that you can get rid of faults and desirable qualities with equal speed. So you see how essential it is to know your dogs thoroughly, and any faulty pup should be destroyed.

Pedigrees are important, but far more important is a knowledge of the good qualities and failings of as many of the individual ancestors as possible. Keep your standards as high as possible, and never mate an exceptional dog to a moderate one, whilst no two dogs with the same fault should ever be mated; it is best to have the good qualities on both sides of the mating. Good to bad will only produce moderate, and excellent to moderate will only produce moderate plus. Aim high, be ruthless with mistakes, and you will produce dogs which you will want to keep, not to sell.

Good food and plenty of it is essential but don't be worried if your Vizsla appears skinny in comparison with other breeds. They are not a breed which runs easily to fat. They should have the appearance of lithe agility, and it is hard enough to get sufficient flesh on them for their ribs not to show, and even harder to keep it there in the shooting season. Despite this they are good and easy feeders, and develop quickly into robust noisy puppies. Generally speaking they are not destructive.

Selecting the right homes for Vizslas is perhaps the most difficult part of the breeding. If the prospective buyer has already owned a Vizsla there is no problem, but when dealing with newcomers to the breed much care must be taken to ensure that they understand fully the temperament likely to develop and the potential working ability in the field it is possible to attain, if a real understanding between dog and handler is achieved.

In every way a Vizsla can be recommended as a family dog. They are excessively clean in their personal habits, almost catlike in the way they will clean themselves and each other. They are demonstratively affectionate and *need* it to be returned. They cannot be kept happily in kennels. They love children and will take considerable teasing from them, although I cannot in any way comprehend parents who do not instil in their infants as much respect for their dog as for themselves. They are good guard dogs in that they bark but do not bite unless in defence of themselves or their property. They are intensely loyal. An example of this was when I was given a bitch of six years old. Her owner had imported her, trained her and bred from her. He then found himself over-dogged and asked me to have her. They arrived from Scotland about teatime and after giving the owner a cup of tea and 'Shian' a run, I put her in a kennel until he had gone. As she seemed to make friends with the four other dogs that lived in the house I decided to have her with us and not sleeping in a kennel, as at that age a change of not only home but also environment would take quite a while to accept. Oddly enough she appeared to be quite content, and ate and slept well. Nevertheless all day, except when she was in the house or up the fields with me, she would sit in the drive ears pricked with an expectant look on her face. As cars and people came and went they were inspected quietly by her and then she would return to her vigil. Shian made it absolutely plain to me that she was quite prepared to stay with us contentedly until her owner returned. It took over a month before she resigned herself to the fact that he was not going to come back and after that she relaxed.

Two years later we met him at a Game Fair, or rather Shian met him. I had taken her with me and was talking to some friends when she suddenly slipped her collar and trotted away. I called to no avail so followed her for about 100 yards through a normal Game Fair crowd until I saw her greet her ex-owner with such ecstasy it was heart breaking. He was not a hard man, but showed no great affection; a pat was as much as she got—or expected. However, she never left his side until 6 o'clock when he said goodbye. I admit I was upset in a way as I had thought she must have been happier with me where she had the comfort of the house, the company of the other dogs, plenty of good food and shooting. But I was wrong. Her loyalty was stronger than the superficial benefits supplied by me. She didn't meet him again. By the time she died at the age of 14 I think I had captured her love and affection. It was certainly something worth having.

Although the three pointer/retriever breeds best known in this country have much in common, they are totally different when considered in careful detail, both in conformation and character. The Vizsla, numerically the smallest still, is perhaps the most elegant with his noble head, lightly boned but substantial frame, swift and effortless movement and the bright sheen of his russet gold coat. He is also probably the most difficult to train successfully to field trial standard owing to the mixture of gentleness and rugged self-will. He is certainly the most effusively affectionate and utterly appealing, even when caught in the most direct act of deliberate disobedience. 'It's not the dogs I mind, it's the people' as a London bus conductor once said to me when I asked him if I could take a dog on his bus. And there lies the crux of the matter. All things can be achieved by the right dog with the right man and this is the most important factor when considering the Vizsla.

20 Field Trials for Retrievers and Spaniels

Field trials for gundogs have been in existence for nearly one hundred years. It was in 1865 that the first event, for pointers and setters was staged at Southill, Bedfordshire, on the estate of Mr W. H. Whitbread. Bird-dogs were then the 'in' dogs of the shooting fraternity because of the system of agriculture, the stage of development of the sporting gun and the conduct of sport generally. It was not until thirty-four years later, in 1899, that field trials for spaniels and retrievers were held. In January of that year the Sporting Spaniel Club ran its first working trials on Mr William Arkwright's Sutton Scarsdale estate, and some time during the following season there was a trial for *retrieving* dogs near Havant, in Hampshire, organised by the same body, along with all-age, puppy and brace or team stakes for spaniels.

The retrieving dog stake is particularly interesting, because of the assortment of breeds which competed. This consisted of one Irish water spaniel, two Clumber spaniels, one Field spaniel, one Curly-Coated retriever and five flatcoated retrievers. All the awards went to retrievers, but this served only to inspire the breed supporters to greater things, and in January 1900 the Spaniel Club held trials at Welbeck, where fourteen dogs competed in the open stake (which did not call for retrieving), ten dogs ran in a stake for retrieving spaniels, and six brace in the brace stake.

In 1900 and 1902 retriever field trials were staged, all the dogs being Flatcoats, and in 1904 the International Gundog League put on a stake which attracted 19 entrants—one Labrador, one Curly-Coated retriever and 17 Flatcoats. In the same year the Kennel Club woke up to its responsibilities and organised trials for retrievers, the all-age stake having twenty entries. Gundog owners were, obviously, beginning to be 'bitten by the bug'.

So it can be said that, from the turn of the century, field trials have proliferated, becoming increasingly popular, not only with the 'big' owners and professionals, but with the dedicated 'private' working gundog fanatics as well. With the exception of the War years (1914–18 and 1939–45), bird-dog trials have been going strong ever since their inception in 1865, and those for the other breeds since around about 1900. However, the pattern has changed with the years, for a number of reasons; these include modern intensive farming methods, the popularity of driven, rather than walked-up, game shooting, and the consequent hand-rearing of game and, above all, general prosperity and the availability of shooting, of all forms, to a much wider public. As a direct result, pointer and setter work declined in popularity—albeit managing to hold its own—and retrievers and spaniels have come into prominence for utilitarian reasons, whilst the pointer-retriever breeds are showing a slow, but steady, increase in favour with all-round shooting men.

All the major gundog breed societies stage one or more field trials each season, whilst many more are run by clubs and societies formed specially for the purpose, all over the British Isles, details of which are obtainable from The Kennel Club, 1–4 Clarges Street, London W1Y 8AB, along with all the relevant rules and regulations governing the conduct of trials.

The object of field trials is—and always has been—to discover the best working strains and the best trainers and handlers, for the improvement of the gundog breeds and *for the benefit of the shooting man* who uses a dog to enhance his sport and conserve game. This latter consideration, regrettably, is often overlooked by the general shooting public, many of whom tend to criticise field trials, usually unjustifiably, for being too artificial and something of a circus performance. Nothing could be farther from the truth, and if only some of the armchair critics would take the trouble to attend a few top stakes, and discuss with handlers the general gundog scene whilst observing the dedication of most of those folk participating in the trials, their eyes would be opened. Furthermore, it is quite obvious that much of the criticism hurled at trials, and trial dogs and handlers, is no more than 'sour grapes'.

I am not for one moment claiming that field trials are perfect, because they are not, never have been and, possibly, never will be. One only has to consider the vast problems of organisation; the difficulties of obtaining enough suitable grounds with an adequate supply of game; the short hours of daylight between October and January, and the human frailties (and differing assessment of gundog work), of judges, coupled with increasing applications for nominations in every type of stake, for all breeds, to appreciate that it is impossible to test the competitors as thoroughly as would be the case during a normal day's shooting.

However, *as far as is possible*, trials are conducted like ordinary shooting days for the breed concerned but, of necessity, the amount of time each dog is operating is limited, and not every contingency can be catered for. Circumstances, inevitably, vary from trial to trial, from day to day and from beat to beat—almost from minute to minute—according to the ground encountered, the game lying thereon, the efficiency of the guns, and the direction of the wind. Without any shadow of doubt, luck plays a very important part in field trials, but a really good dog will come to the top in the end if given a sufficient number of runs. The chief difficulty

78. H.M. The Queen's yellow Labrador Sandringham Sydney retrieving from across water at the Kennel Club's Centenary Retriever Trials at Sandringham in October 1973.
Peter Moxon

79. 'Casting out'. Keith Erlandson in action at the Spaniel Championship at Blenheim Park in January 1969.

today is to obtain nominations in enough stakes to prove a dog's worth.

One very important aspect of gundog work which field trials cannot possibly discover is *stamina*, because if 12 to 16 dogs have to be tested in about six hours, it is obvious that none is going to be able to have a long enough 'down' under the judges to prove its staying power. It is difficult to see how this imperfection can be overcome without considerably reducing the number of entries in, and extending the duration of, every stake, something which, in present-day circumstances, is quite impracticable. However, too much can be made of this point, because most, if not all, field trial handlers are practical shooting people who would not dream of burdening themselves with dogs which could not stand up to a hard day's work in the field.

Field trials are great fun, both from the practical dog work and the social points of view. The vast majority of field trial supporters, in all breeds, are very good sports, always willing to help the other chap with advice. Even more important, in my opinion, they manage to retain a sense both of proportion and humour; they try hard not to take themselves, and their dogs, too seriously. Of course, there are exceptions, and a minority of both professional and private owners and handlers have, over the last few years, become so commercial-minded and 'kennel blind' that they just cannot take defeat, and tend to make life miserable for everyone after a stake by complaining about the ground, the guns, the game supply or, more frequently, the incompetence of the judges. Experienced sporting triallers have learned to live with such folk, and to see their funny side, thoroughly enjoying the 'inquests' that inevitably occur at the headquarters hotel on the evening after a trial, where a great deal of leg-pulling and 'mickey-taking' goes on, just as it does in the spectators' line during the course of a stake.

Newcomers to trials are sometimes put off by what I term the 'lunatic fringe' of moaners, but they should persevere, and will soon learn that these people should not be taken too seriously. The hardcore of triallers are wonderful people who will bend over backwards to help and encourage beginners.

THE CONDUCT OF FIELD TRIALS

Whilst a gundog trial is conducted as nearly as possible in the same manner as an ordinary shooting day, there must, obviously, be differences, because the emphasis is upon *the work of the dogs* and not the quality of the shooting or the size of the bag. Beats have to be taken very slowly, and more time allowed for the collection of game, than would normally be the case, when pickers-up would do the job after the guns had moved on. The judges have to be given time to make notes after the performance of each competitor, and the field, especially the gallery of spectators, organised properly in order that they are in no physical danger and will not interfere with the dog work. Where retrieving dogs are concerned, every bird which hits the ground must be made use of, and this often means that a dog will be sent into unshot ground of the next beat for 'long' birds—dead or running, thus disturbing and wasting game, something which, on a normal shooting day, would certainly not be allowed.

Field trials for retrievers are held over one or two days, for either 12 or 24 dogs respectively. Game can be walked-up or driven, or a combination of the two systems, according to the ground and the wishes of the host. There may be three judges, with two dogs and two guns allotted to each, or four judges, working in pairs, with two dogs under each and with two, or maybe three, guns shooting for each pair. The chief steward keeps the judges supplied with dogs, which are always numbered from the right of the line, and tested in numerical order during the first round of the stake, and each judge, or pair of judges, has a steward acting for him. The line of spectators is controlled by a steward carrying a flag, who endeavours to position them so that they have a reasonable view of proceedings, yet are safe from the guns and unlikely to interfere with the work of the dogs.

Each handler wears a numbered armband, to coincide with the number his dog drew in the ballot previous to the trial. This also appears in the

80. An English Springer completing the water test at the Spaniel Championship in 1973.
Peter Moxon

official catalogue, or 'card', along with details of owner, breeder and sire and dam and date of birth of the dog. Usually, each judge is accompanied by a steward carrying a number board, which is lofted for the spectators to see when a dog is put out to work. In the first round of a stake, the judges endeavour to use the birds shot by their own guns and to give every dog a first retrieve, or at least the *opportunity* to retrieve a bird which it has (or should have) marked, although, of course, some retrieves may be of the 'blind' variety owing to circumstances.

If the first dog fails, the second is called in to have a try, and given a little more time to make good. If this also fails to collect, and the judge considers the bird 'gettable', he may then offer it to his colleagues, who try their dogs in numerical order, but only the dogs actually in the line at the time are tested. If all fail, and unless it is an obvious runner, the judges may go forward and look for the bird themselves and, if it is found, these dogs are considerably downgraded to the point of elimination from the stake.

Unless a dog is guilty of an 'eliminating fault', such as unsteadiness, hard mouth or whining, it must come under two judges before being discarded from the stake. This means that, at a trial where three judges are officiating, most dogs get at least two runs. The four-judge system, mentioned above (which, incidentally, has operated at the annual International Gundog League championship stake for several seasons past) saves time, ground and game in this respect in the case of less impressive dogs. The judges take notes of the performance of each dog, conferring at the conclusion of each round and grading competitors 'A', 'B' or 'C' (with plus or minus symbols where appropriate). Thus, by the process of elimination, they decide which dogs should be seen again and, ultimately, compete in a 'run-off' between the top dogs, if this is deemed necessary.

It sometimes happens that a particular dog works so brilliantly throughout the body of the stake that it is the obvious winner, in which case the judges may keep it 'on ice' and run dogs off for the other placements. Or many dogs may go wrong and the top competitors just place themselves in the first two or three rounds, thus removing the necessity for a run-off. What usually happens is that, at the conclusion of the second round, there are five or six 'A' dogs and several in the 'B' category worthy of consideration, so these are tried again, with all judges watching them together from the centre of the line, generally testing the 'B' dogs first, so giving them the opportunity to elevate their positions. A final run-off of the 'A' dogs decides the placements, by the process of elimination. Whenever possible, a water test and a drive is included, if these have not been practical in the body of the stake.

SPANIEL TRIALS

Much of what I have written about retriever trials applies also to spaniel events, especially as regards general organisation and the classifica-

81. Line for a drive at a Retriever field trial in Scotland.

tion of the dogs' worthiness by the judges. However, there are, naturally, important differences, because spaniels are *questing* for game as well as retrieving it, the major *role* of the spaniel being to find and flush unshot game.

In the spaniel stakes, up to 16 dogs can be tested in a day, or 32 over two days, under two judges and a 'floating' referee. Again, numbering from the right of the line, each judge has one dog down under him at a time, with, usually, two guns allotted to each judge, as well as the appropriate stewards. The dogs are expected to quarter the ground within gunshot range and to retrieve game found and flushed by them, no matter which gun in line actually shoots it. They must prove steady to flush, shot and fall (preferably dropping to rise and shot), and retrieve tenderly to hand when called upon to do so.

If a dog fails to find, the judge offers it to his colleague who then tries the dog down under him, when an 'eye-wipe' may ensue, which is duly noted. If both dogs fail, the judges will themselves try to find the game (unless an obvious 'runner') and, if they succeed, the dogs tried will be eliminated from the stake. The judges work through the card, the right-hand judge dealing with the odd and the left-hand man with the even numbers, subsequently changing over, so that every dog is seen by both judges, unless discarded for an eliminating fault in its first run. The unfortunate referee, walking in the centre of the line, and dodging about from time to time, tries to see as much as he can of all that goes on, but takes no part in the judging *unless* called upon to offer an opinion on the work of a dog about which the judges cannot agree.

As in the case of retrievers, a run-off between the top dogs may or may not be necessary, according to the quality of the work in the body of the stake, and the winner may be kept 'on ice' if agreed upon by both judges. However, at *all* trials, most judges prefer to hold a convincing run-off between the finalists, thus making them prove their placements and letting the handlers and spectating public see that justice has been done!

What are field trial judges looking for? First and foremost, *gamefinding ability*. Control and 'handability', 'gun-sense', marking ability, perseverance, tender mouth, courage in cover (and water if tested), and general method of ground treatment, which is especially important in the case of spaniels, are all taken into consideration. In the final analysis, the other so-called 'finer points' of gundog work can have an important bearing on judicial decisions, these being pace and style, clean (sitting or standing) delivery of game, and initiative under difficult circumstances. It has been said that a field trial dog is only a competent shooting companion more highly trained and 'polished'. Our best judges bear this in mind and try to imagine that they are themselves shooting over the dogs which are running under them. In this way they obtain an *overall* picture of each dog's performance, and are not unduly swayed by a single brilliant piece of work, such as the collection of a difficult runner, for instance, if, in other aspects, serious defects show up.

Judging field trials can be a pretty thankless task, albeit a very interesting one. Handlers and spectators often criticise decisions quite unjustly, usually because they (unlike the judges *in consultation*) have been quite unable to evaluate *all* the work of every dog competing, and sometimes go away from a trial with completely the wrong impression. However, in my own experience, extending over nearly thirty years, the vast majority of judges are 'straight' and the few injustices which have been done were due to human error (sometimes sheer ignorance!) rather than bias or deliberate unfairness.

The ideal judge, if there be such a paragon, must have the ability to concentrate and understand the type of dog he is judging; he needs the patience of Job, the hide of a rhinosceros, a sense of humour, and the ability to stick to the courage of his convictions. So long as he can drive home convinced that he has been just and found the best dogs *on the day's showing*, why should he worry?

Handling a dog in trials is at first a nerve-racking experience, until the novice gains experience and confidence. Some people, indeed, still suffer from trial nerves after years in the game, and readily admit to it, yet still keep coming back for more! Nervous tension in the handler is inclined to rub off on his dog, with unfortunate results, and countless cases of so-called 'bad luck' can be directly attributed to the handler 'flapping' and grossly *over-handling*, doing all sorts of stupid things which he would never do on an ordinary shooting day. Our most successful trial handlers appear to have nerves of steel and great singleness of purpose and powers of concentration. But even some of these do, occasionally, show signs of stress, especially in a hot corner or during the run-off, when they are within sight of victory.

I hope that I have written sufficient to whet the appetite of owners of working spaniels and retrievers at least to support and watch, if not actually compete in, field trials for the breed of their choice. The trial 'bug', once caught, is difficult, if not impossible, to shake off if you are genuinely interested in practical gundog work and are a true sportsman—

or *sportswoman*, for that matter, because, over the last twenty years the female element in trials, for all breeds, has come very much to the fore.

Trials are great levellers, and in the field everyone is on the same level with no class or sex discrimination as in so many sports. It is true that, over the last few years, a certain amount of commercialism and 'gamesmanship' has crept into trials, mainly due to increasing cost of participation and the very natural desire of even the amateur section of the fraternity to recoup expenses by sales of trained dogs, breeding and stud fees. However, this only really increases the challenge of the game, and in what other sport does the novice amateur have the opportunity to compete against the top stars on equal terms.

Glossary of Field Trial Terms

Run or *Down:* The period when handler and dog are operating under judge's orders.

Failure: When a dog fails to find and retrieve dead or wounded game.

Eye-Wipe: The collection of a retrieve after another dog has failed.

Hard Mouth: The damage of game by a retrieving dog.

Find: The locating of game by a questing dog.

Blinking: Ignoring game which should be flushed or retrieved.

Running-In: Moving forward without handler's orders, at flush, fall or shot.

Giving Tongue: Barking or yapping whilst working.

Whickering, *Squeaking:* Whining at a drive or in the line when walking-up.

Runner: A wounded bird which has moved on.

Switching Birds: Dropping a bird being carried and picking up another.

Marking: The ability to watch and pinpoint the fall of game.

Pegging: Catching and picking up unshot game.

Style: Attractive movement and tail action and speedy, eye-catching manner of working.

Pattern, *Ground Treatment:* The manner in which a dog quests its search area methodically (and within gunshot in the case of spaniels).

Blind Retrieve: A retrieve which the dog has been unable to mark.

Taking a Line: Following the scent trail of a wounded bird.

Flush: Putting up game during quest.

On Ice: Term applied to the dog which has performed so well during the body of a stake and proves the winner without participation in the run-off.

Run-Off: The concluding rounds of a stake when the judges, walking close together, compare the working of the highest-rated dogs of the previous rounds to decide the final placements.

Kennel Blind: Term applied to an owner or handler who will not admit the worth of any dogs except his own.

21 Working Tests

Over the past 20 years, gundog training has become increasingly popular with the 'ordinary' owners of the working breeds, whether their dogs are intended simply for shooting over, as competitors at field trials, or perhaps both. It was inevitable, therefore, that numerous breed and trial societies should jump upon the bandwagon to encourage this interest by organising training classes, under the guidance of competent instructors, to teach novice owners and handlers just what gundog education is all about. Their aim was also to achieve a reasonable standard of training, step by step, in a progression of exercises designed to 'bring the dogs out' so that they would be capable of operating in the field in an efficient and bag-filling manner.

I can justifiably claim to be one of the pioneers of this movement, because in 1950 I was asked to act as instructor at training classes for retrievers, organised by the United Retriever Club, in Kent. Here we found enthusiasm running very high. Such was the thirst for knowledge that classes of between ten and twenty dogs, of various ages and breeds, and in different stages of training, were not uncommon. We quickly learned that handlers and dogs alike would have to be divided into separate sections—beginners, intermediate and advanced and also that, at the end of the series of monthly training sessions, some sort of 'passing out' examination was both necessary and desirable.

Thus were born gundog working tests, as we now know them, and both they and training classes, have proliferated throughout the country. So much so that they have almost become a sport in their own right, not always, let it be noted, to the approval of the working gundog fraternity (especially field trial fanatics) in general.

However, I am quite convinced that both training classes and working test have done, *and are still doing*, a great service. Not only for the various breeds of gundog, but also for average owners and the shooting man, by fostering interest in, and the efficient training of, dogs by people who would not otherwise have access to the necessary facilities, or the knowledge to set about the job.

Without question, tests provide opportunities, under ideal conditions, for gundog folk to gather during the close season, to compare notes and dogs and, in general, keep things swinging until the shooting and field trial season comes round. Furthermore, they undoubtedly feed into the field trial pipeline a continuous stream of newcomers, without which the sport would die a natural death. So, if we accept the premise that field trials are of benefit to the gundog breeds and the shooting man (by discovering the best strains of dogs and the best handlers), we must also agree that classes and tests make a very valuable contribution to the practical working gundog scene, even if there are valid reservations on the part of some gundog folk, objections with which I will deal later.

Gundog tests must not be confused with field trials proper. They consist, for the most part, of a series of artificial exercises, usually held during the spring and summer, on limited ground, with all the competitors performing over the same course and judged by one, or sometimes two, experts. As a rule, but not always, dummies are used for retrieving, instead of the real thing, both on land and in water, and very seldom is live game—or even natural game scent—encountered.

The majority of tests are designed to demonstrate that the dogs have been thoroughly 'hand-trained' by the preliminary educational course, are under control, 'handy', and are capable of questing, and retrieving from land, thick cover and water; they must be able to negotiate fences and other obstacles when called upon to do so and, of course, must not be gunshy. In other words, the dogs are at that stage of training when the practical handler considers them ready to embark upon genuine field work on natural game. By their very nature, tests cannot *usually* offer any proof of tenderness of mouth, efficiency of nose on the line of a runner, of steadiness to rising game or, for that matter, steadiness to a heavy downfall of shot game in a hot corner at a drive.

In some of the more ambitious tests organised at Game Fairs and the like, pigeons may be released from traps concealed in cover, dead birds (usually pigeons) used for retrieving, and the dogs made to pass wire pens containing rabbits, in order more nearly to simulate natural conditions and provide temptation. But these are the exception rather than the rule, and the usual club-organised test is a much simpler affair.

Convincing tests for retrievers are comparatively easy to arrange; those for spaniels considerably more difficult, whilst to simulate genuine bird-dog (pointer and setter) work by artificial means is well nigh impossible for reasons which will be obvious to anyone with experience of the different types of gundog and the type of work they are required to do. It has been said that 'you can train a retriever whilst feeding the chickens', which is a bit of an exaggeration but truthfully indicates that you do not need to have too much ground and time for the job if the dog is intened simply as a no-slip worker. Spaniel work calls for a greater area of ground because it must have sufficient room to quest, and also cover to investigate. Bird-dogs, if they are to be utilised in the proper capacity, require a very large

82. Flatcoat Retriever negotiating a fence at a working test.
Peter Moxon

area of ground indeed upon which to range and, furthermore, the real thing (even if only pigeons concealed in cages) to wind and point. True, you can train all three types of dog (*for basic obedience* in the yard or on the lawn, and also hold very limited *novice* tests in a confined area, but advanced training and more ambitious tests demand the right ground for the job, including cover, water and some form of jump.

Organisers of tests also have another consideration to take into account —and a very important one in most cases—and this is that the gallery of spectators shall have a good view of proceedings. So the course has to be laid out with this in mind, more especially when the event is staged as a demonstration, attraction and public relations effort at a game fair or country show. Unfortunately, few grounds offered prove ideal in every respect, and the backroom boys and girls who do all the hard work of general organisation often have a hard time dealing with complaints from handlers and spectators, because the layout of the event is not completely suitable for both.

Because they are unavoidably artificial in nature, tests are *usually*—and I emphasise the word—best performed by young dogs up to eighteen months or two years of age. Once a gundog has had a taste of natural work on game and been shot over or competed in trials, it tends to lose interest in dummies, set courses and limited ground, which it rightly considers 'kid stuff', boring in the extreme and quite beneath its dignity, especially in unseasonable weather in spring and summer. However, there are exceptions, and I know of (and have owned) many mature dogs which work with enthusiasm, pace and style on scentless, and perhaps fouled, and limited ground, going through all the motions of their earlier education, and in front of large noisy crowds. These are the 'naturals' (and, I suspect, extroverts) of the canine race, and are a joy to own and handle.

I must add that sensible handling has a great deal to do with this (as, of course, does breeding) because work—especially retrieving—for a gundog should always be *fun* and a privilege rather than a chore.

One of the greatest failings exhibited by novice and over-enthusiastic trainers is to overdo the monotonous early discipline and dummy work, especially under unfavourable conditions of weather and scenting, to the point of boredom for the pupil. Another weakness displayed by numerous test (and, for that matter, field trial) handlers, is 'brushing up', 'conditioning', call it what you will, their dogs immediately prior to the competition. They will often give them such a 'going over' the day before, tiring them out, mentally and physically, that they are completely unable to give of their best on the day.

I firmly believe that a fresh dog will function far better than a stale one, and it is a well-known fact that a tired dog is much more likely to go wrong, especially as regards steadiness, than a lively and rested one. From my observations of tests and trials over the last twenty-five years, it is patently obvious that many of the more disappointing competitors have been subjected to too much attention immediately before the event. By all means 'sharpen a dog up' on command and whistle *for a few minutes on the morning of the test or trial*, after having conditioned it generally earlier in the week, but avoid like the plague the understandable temptation to give it the 'full treatment' on the days immediately before the competition. This will only defeat its own object, as I know to my cost.

Handling a dog at tests is just as nerve-racking as it is at trials, if not more so, because of the usually large, critical and very close gallery of spectators. The handler's nerves inevitably rub off upon the dog, and there is just nothing that can be done about it. The judge, or judges, are breathing down your neck even more so than at field trials and, because each competitor has almost exactly the same tasks to perform over the set

83. Celebrated handler, Bob Baldwin demonstrating a clean delivery of the dummy at a working test.
Peter Moxon

course, the handler is understandably over-anxious about slickness of performance and the time factor, often of supreme importance in these artificial events. The great thing is to forget about the audience and the judges (if you can!) and concentrate upon your dog, above all regarding the whole thing as a sport and a bit of fun. Don't panic if mistakes are made. Competitors in field sports tend to take themselves too seriously, and to worry unduly when things go wrong.

Field sportsmen (and women) are a notoriously conservative (and often prickly!) fraternity. Field trials proper have always—wrongly in my opinion—been the subject of severe criticism from shooting men, the very people they are specifically designed to help. Gundog working tests and, indeed, even training classes, have, in their turn, been slammed by field trial purists for giving owners and handlers a false conception of the worth of their dogs for more advanced competition and work in the shooting field.

Now, it is perfectly true that the successful test dog, pure and simple, is not necessarily going to prove an adequate shooting companion, let alone a star turn at field trials. Unless, and until, it has had experience on the real thing under practical shooting conditions it is an unknown quantity.

Tests, ideally, are only a beginning, and a means to an end, and not an end in themselves. But, unfortunately, some owners have been so carried away by moderate successes in artificial competitions that they have been entering their dogs for field trials—even the major events—*without first giving them more advanced work and experience in the field.* They not only make complete fools of themselves and their dogs, but also waste nominations in over-subscribed events, which could have been taken up by more worthy contestants. Worse still, these misguided folk have lowered the standard of field trials and given a nasty shock to spectators in general, and the hosts of the trials in particular. This, to say the least, is extremely bad public relations at a time when grounds for trials are becoming increasingly hard to find, to say nothing of the difficulty of obtaining a nomination in even puppy and non-winner trial stakes.

I believe that it is time that gundog fanatics got their priorities right. In the first place they must decide whether they wish to go all the way and make their dogs good enough to shoot over and/or compete in field trials, in which case they have simply *got* to have access to game in its natural habitat, shooting and, perhaps, picking-up. If, on the other hand, they are content simply to enjoy training classes and working tests as a sport in their own right, and, through lack of ambition, funds or time, or a combination of circumstances they are unable to progress, they should not make life more difficult than it already is for the serious trial competitors, by wasting nominations and putting up a poor show in front of the very people who have the future of the working breeds in their hands.

Tests and training classes are great fun and not only have done, but

84. Spaniels under 'steadiness to rabbit' education at a training class.

can still do, a valuable service for the working breeds of gundog, their owners, handlers and breeders, and the ordinary shooting man, *so long as they are kept in perspective*. Let us keep them this way, and discourage owners and handlers from getting an inflated idea of the value of their dogs in top competitions at trials.

22 Picking-up

Picking-up is a job and sport in which male and female, old and young, with dogs of many shapes and sizes, ages and ability, can take part. It ranges from going out with a semi-trained dog, and one shooter with only a hard-earned pigeon or rabbit to show for one's endeavours at the end of the day, to forming part of a high-powered shoot with over 300 birds in the bag and four to five experienced pickers-up with a team of top class dogs, working silently and hard, late into the evening, long after the guns have gone home oblivious of your existence! There are many niches along the line into which you and your dog can fit. For those who do not shoot —like myself—and are trying to prepare a dog for the shooting field and trials, the experience is well nigh essential.

TYPE OF DOG

Choose any of the working shooting breeds—labradors, retrievers, or spaniels—if you are going in for shooting seriously. These have been bred for work for generations, and have natural retrieving instincts with good noses and trainable temperaments. Beware of the dual-purpose show dog, unless you actually see it work or, if it is a puppy, can watch *both* parents working in the field.

On the other hand, one can have great fun with a mongrel and I have seen several excellent workers that were not a recognised breed of shooting dog, but have given reasonable service to their owners in one way or another. One such non-shooting breed was a large standard Poodle (Poodles used to be worked in France as shooting dogs) called Clem (after the late Lord Attlee) owned by Lord Butler, which gave him yeoman service over the years.

We have had a Springer Spaniel, working Welsh Collie and working rough-coated hunt terrier. The two last were so bored being left behind on shooting days, that I started taking them out beating on days around the farm. They were obedient and extremely useful hunting in sugar-beet. Both would find the shot birds and stand over them until the Spaniel collected them. The Collie got so bored by this that one day she picked up her bird and, hotly pursued by the Spaniel, brought it to me. I thought it might be crushed to bits as she was a wonderful ratter but there wasn't a feather out of place, and from then on she retrieved—albeit by foot or wing! The terrier also got bored by the Spaniel retrieving and he'd stand, hackles up and teeth bared, until I came up and rescued him and the bird.

Another time I witnessed a most charming sight. The guns backed onto a lake and in the line stood the following four breeds; a black labrador, German short-haired pointer, golden retriever and my springer spaniel. All the guns dropped birds roughly the same distance way out in the water, and by coincidence we all sent our dogs to retrieve at the same time, each retrieving his own bird and swimming back in line abreast—and no one had a camera!

STANDARD OF TRAINING

Before you take your dog anywhere to work in public, training is necessary to achieve the following qualities:

1. Implicit obedience at all times to voice and/or whistle—preferably whistle.
2. Proven soft mouth.
3. Non-fighter and non-whiner.
4. Extra refinements: controlled initiative; ability to take directions; ability to jump; good delivery and work with all types of game.

As far as obedience is concerned he must *come back or sit at once* when ordered; he might, for example, be on the edge of a covert about to be shot or on the point of chasing a runner in front of guns still shooting. He must *leave* when ordered, e.g. (i) If about to 'peg' or pick up un-shot birds; (ii) If you have a strong runner and dead bird, you wish him to get the runner first, and he goes to the dead bird; (iii) Or anything else he is about to do that you don't want him to do, such as taking a bird out of another dog's mouth!

Steadiness to shot, flush, and other dogs is an essential corollary to obedience training, and will have been instilled from puppyhood.

Lack of jealousy. Dogs are born jealous and one has to help them get over this (although I have never seen it quoted in any book on dog training). I give them a course to help them curb their feelings by the following methods: (i) make them sit while another dog goes and retrieves near them; (ii) pat and caress another dog near them; (iii) I walk off with another dog, leaving them sitting; (iv) another dog hunts around them.

Jealousy leads to all sorts of troubles, such as taking a bird from another dog's mouth. I had a very promising puppy made hard-mouthed for ever by this. While in Scotland, having marked a grouse down, I let her go to pick it up. It was alive, and she found and gathered it well. As she was galloping back, a huge Labrador rushed up, knocked her down, fought her, took her grouse from her and left her with a bleeding nose—she had to be rushed to the vet to have it cauterised and never an apology from the Labrador owner! Afterwards if she saw a black dog anywhere near

85. A good day's work.
Roy Shaw

her she would crush her birds. Another mark of jealousy is hunting beside another dog that is hunting, instead of keeping to his own territory. Both merely get faster and faster, stop using their noses and gallop about.

Whining is hereditary, incurable, and catching, but also can be triggered off by injudicious handling, such as having a puppy too close to too much game too soon, or by jealousy again, caused perhaps by seeing the next-door dog always running in to shot and getting all the retrieves or being next to a winning dog. In the last two instances retire rapidly out of sight of your neighbour's dog!

I mentioned earlier 'Obedience to voice, preferably whistle'. The human voice carries a long way and will disturb game for quite a distance *especially* if you are upwind (i.e. with the wind blowing from you to the unshot cover). Nothing makes pheasants run quicker than hearing you calling to your dog, but a whistle, your own or artificial, has much less effect on game, carries farther and you'll be far less tired by the end of

86. Waiting their turn at the next drive.
Roy Shaw

the day. One needs only three basic notes: 'pip' for sit; 'pip-pip' for turn, or look at me; several 'pip-pip-pips' to indicate return at all costs. You should practice until perfect at home, behind a hedge or screen where the dog does not think you can see him. Then you will have a chance of him stopping when you are genuinely out of sight in cover.

Do not take your dog to pick up at other people's shoots if he has a hard mouth—very many people have no idea if their dogs have a hard mouth or not. This is easy to detect if you know how; run your fingers down each side of the backbone of the retrieved bird and if you find one side (or both) is caved in, or you can feel broken ribs, be suspicious! He *may* have had a hard tussle through the brambles; this is understandable and forgiveable but if it happens again and again, I fear there is no cure for him. These birds are worth nothing in the market, won't hang well and can only be used in stews.

If you want to test a trained puppy, before he begins his first season, use a dummy made from a plastic bottle in an old sock and examine the bottle from time to time to see if it is punctured. If you have a dog with a proven hard mouth *please do not breed from him or her*. It is mainly hereditary and trainers have enough hazards to contend with, without being asked to spend hours and weeks on a dog's education, only to find, when the season starts, that it has inherited a hard mouth.

It is extremely selfish to bring a fighting dog into the shooting field, although the owners usually have an excuse for it, 'He doesn't really mean it' etc., they never seem to realise what unpleasantness they cause and ten-to-one if your dog is attacked, he will be spoiling for a fight himself the next time. I have seen two accidents of different types caused by this. In the first, the guns and dogs were on a tractor-drawn trailer, when a fight ensued between two huge Labs. and one of the men *plus his gun* was tipped over the side—result, a strained arm and a dented barrel. The second, was the incident I recounted to you earlier, under 'Jealousy', which turned my light-mouthed bitch hard-mouthed for ever.

ABILITY TO TAKE DIRECTIONS

A dog must be willing to take directions from you and not gallop vaguely about in the hope of finding game, especially if it is near the next drive. If he will take hand signals from you he can then be relied upon to go right to the edge of the next cover to collect a dead bird, but not go in. You should be able to direct him through a fence, or straight across a field, not hunting his way out; then, at a given signal from you, he must take over and work out his own salvation. All this saves time and energy, but it means a great deal of patient training at home, especially with spaniels, whose idea of heaven is to hunt their way out.

A word here about wind. A lot of shooting men with dogs do not take this sufficiently into consideration. If you are going to direct your dog out to a blind retrieve, note from where the wind is blowing and send your dog out *downwind* of the bird. Dogs automatically turn *into the wind* and stand a far better chance of finding the bird. If you send him *upwind* of the bird he may miss smelling it by six inches and be classed as having no nose—whose fault? Don't come to the conclusion from the above that you must always handle your dog out. There will be plenty of times, such as when hunting up after a drive, when you never touch your dog for half an hour on end and it is entirely on its own. A dog is not fully trained, if it will not take directions and orders, for it may spoil the whole of the next drive.

Teach your dog how to *sit and stay put* (out of sight too!). This can be extremely useful. You can put it to endless uses. Make your dog a stop in a ditch or gateway, or between guns in a thick wood, or even as a canine barrier, where birds are running across from one section of wood to the other. This always looks very spectacular if you place three or four dogs at twenty-yard intervals down a ride, but dead easy if they know they should not move.

Here I will tell a story against myself—an occasion when I wished the ground would open up and swallow me and my dog! The wood was a long stretch, taken in two halves, the second half containing the reared birds. Just behind the guns, after the first drive, was a very thick wired-in ten-year old plantation full of brambles. I was away on the left and two labradors were trying to find three running cocks in the thick, but failed. There was a shout for the spaniels, so over I went and with permission from the keeper collected two of the three quite quickly. I was then told to come through with the beaters but my two-year old bitch, Widgeon, got onto the line of the third cock and was seen by us all several times going *away* from the next drive. Then someone said they had seen her going forward, but whistle though I might, there was no sign of Widgeon. With a sinking heart, I followed in line, knowing that there was a high wire fence between Widgeon and the reared birds (I hoped!). When we came to the wire I found an old Suffolk stop. 'Had he seen my spaniel?' 'Yes, missie, she near cot that thar cock. They wur balanced on't wire together, but your little dawg she got wholly cot up, so the cock he popped down and ran off yonder'. At that, all hell was let loose as Widgeon lost her head amongst 300 first-time-over pheasants—they flew out of the wood in every direction! 'Catch that dog', 'Shoot that b.... dog', 'Where's its owner?' rang out. Nothing could I do, until one of the beaters did a rugger tackle on her as she passed and I miserably slunk back to the car. After it was all over, the guns passed a message to me to come and lunch with them, but I could not face them. I had four letters of condolence from them the following day, so I was forgiven *and* asked to bring her again, as she had done yeoman service for them before—and since!

ACQUIRING A PICKING-UP JOB

(a) Ask your shooting friends if you can come; (b) go and see the local headkeepers. (Take a reference as to your dogs' capabilities from a friend, if needed.) Offer your services on the day or perhaps the day afterwards, when some shoots go through the woods again and hunt for anything left

87. It is essential that the dog should be capable of working with all types of game.
D. N. Dalton

from the previous day; (c) advertise.

If you are just with a couple of guns with their dogs and your dog is young, beware it doesn't pick up bad habits, such as watching (and copying) other dogs running into shot, or having too much hedgerow work. I never let my spaniels work hedgerows in public until Christmas of their first year—it tends to make them either go too far ahead if they are fast workers, or, if they stay with the shot, to potter. So you can't win either way! This is a job for an older dog—unless the guns are friends and help you train your dog. On the other hand, on an average shoot, you may be asked to brush in some fences or stubbles, act as a stop, do some beating, or be off behind the guns to pick up, rushing back again to beat in the next drive, then back behind the guns again to pick up, and so on through the day. On a high-class shoot you should have one, or a team, of really well-trained reliable dogs. They will have to work together, hard and fast, often long after the guns have gone home.

CLOTHES AND GADGETS

Wear sensible thornproof, waterproof clothes, consisting of trousers, gumboots, and hat with a brim. Eschew bright colours. (Incidentally, it is surprising how a hatless face shows up.) It beats me how often one sees guns at high class shoots, or in the butts, standing bareheaded, hoping the birds will come over. If one stands behind them, one can see the birds swerving away. Carry a game bag (ladies can get a lightweight one with external string partition); a stout stick, or shooting stick, for dispatching live birds and a yard of blind-cord with a loop at each end, is useful for carrying extra game over your shoulder—it also acts as an emergency dog lead. I always have in my pocket a couple of polythene bags, nine to twelve inches long by eight to nine inches wide and a couple of rubber bands. These, pulled on over your gloves and held in place by the rubber bands round your wrists, keep your hands warm and dry in wet weather—a tip I gleaned from a well-known field trial secretary in Scotland. Have in your pocket a penknife and small finger bandage in a polythene bag. Several times I have been called on to use this to stop bleeding on dog, beater or gun or even to tie up a gate. In the car have a bowl and water, and in very hot weather in Scotland in August I take a plastic-lined dumpy vacuum flask filled with water in my game bag. I found I became extremely popular with all the other canines on the hill, and was always having to fill up from the Land-Rover between drives!

Here is a list of a few more things I carry in my veterinary bag which is always in the car; they are nearly all just as good for humans too: anti-midge cream; assorted plasters; lanolin; antiseptic cream; some sort of soothing eye cream (I use Betsolan) for the frequent bangs and scratches dogs receive to their eyes in heavy cover; a packet of Johnson's cotton wool buds—excellent for digging bits out of eyes, cleaning ears and wounds, etc. and a 3 in. wide roll of elastoplast for bad tears. A large chamois leather is excellent for drying dogs down before putting on their drying rugs. Each dog has a rug (these used to be horse's saddle cloths) which tie across their chests and round their middles.

If you have an estate car, do have a rear washer, heater and wiper unit fitted. This is a very worthwhile extra. I carry two sacks of straw, keeping one dry for the homeward journey if it is a wet day. An old Lilo is very useful too, I find, to put over the sacks to keep them dry if one has several dogs. It is not as cold for them to lie on as a plastic bag.

You have now trained your dog, received your invitation to pick up at a high class shoot, collected your clothes and stocked up your car—off you go . . .

ON THE DAY

Arrive at least fifteen minutes before time, having given your dog a good five minutes run before your start. See the keeper and find out the beats of the day, note any return drives so that you do not pick up behind the second drive before the first one is done. Ask to keep your own car with you, if possible, as you may not be able to move on with the guns. If it is a big shoot, there may be about four pickers up. Find which one is in charge and he will arrange where you are to go, who stays back to pick up the last birds, and who goes on with the guns in order not to miss the first few birds to be marked at the next stand. Do not expect to be given the best places as a new boy until you have proved your worth! If you go on with the guns, go to the far end of the line so the late pickers up can slip in behind the near guns.

See that you are definitely out of shot—if you get peppered, it will be your own fault. The guns must not be expected to think about you or see you moving out of the corner of their eyes. They have to concentrate hard and it is very disconcerting to have to remember your position and then break their swing to avoid you. A good picker-up's maxim should be 'neither seen nor heard'. If you cannot get out of shot, then you must explain to a gun without a dog and ask him if you can stand beside him. But with ingenuity this should seldom be necessary as you can usually find a tree or ditch for cover. Either way, remember the old story of the centenarian who attributed his great age to plenty of beer and 'baccy and the fact that he always fell flat on his face when the cry of 'woodcock' went up!

If you are in a wood stand at least a hundred yards behind the line, though two hundred is better. Find a clearing in the trees and watch for birds with legs down or unlevel wings. You might even stand at the far end of the wood, for wounded birds usually run as far as they can in cover, away from the noise and then tuck in; you will pick up a lot in the brambles around the perimeter. If you are on open ground or heath land, two hundred to three hundred yards is better, making yourself as inconspicuous and still as possible. A wounded bird will often set its wings

88. Willingness to jump although blinded by a wing over the eyes.
Gosling

and glide until it thinks it is safe and then land; you can then mark and pick it up, but if you move it may continue to glide out of sight and be lost.

There is another good reason for standing so far back, too. You will be facing the guns marking and picking but if you are only 25 to 50 yards behind, you will always be turning round to mark and will miss the next bird down.

After the drive, work your way forward to the guns, putting the birds down at the pegs of the guns that shot them. The guns can tell you more accurately how many more are left to pick. Nothing is more irritating to the guns if they have no idea who has picked what. It should be a golden rule that *all birds should be put by the pegs* to be checked by the guns. A great deal of time is wasted looking for birds already picked by beaters or pickers up who do not say where they came from. *Never, never* pick up within range of a gun with his own dog unless he asks you to. It is most irritating for a gun with a dog to find that pickers up and beaters have swept through and his own dog has nothing left to do.

You will soon discover two things. Firstly, to distinguish the guns worth standing behind, and who are reliable markers, from the super-optimists who think they've hit everything. 'Just slip over to collect that cock by the far fence on the right and after that another bird stone dead just half a mile to the left.' Phew! You know they have flown off, unscathed, but you must show willing! Secondly, there are the pickers up, few in number I am glad to say, who stay near the guns to impress, bags bulging with all the easy picked birds; they finish first in each drive, never mark a far-off wounded bird, are off for lunch with the beaters and away first in the evening. But beware, don't try to be too clever as I was once, quite by mistake. The neighbouring keepers, who were picking up with me, were very friendly, but they packed up for lunch and I usually carried on if I thought there was something down. This time there had been a return drive, plus a sideways one, with picking up only done at the end of the last one, so there was a great deal to do. I had four dogs and by the end

89. Well-trained spaniels will wait patiently.
Gosling

prayed they would not bring me another bird. I had picked 17 wounded birds, mostly runners. Returning triumphant, the headkeeper was exceedingly pleased but I lost, for a time, the friendship of the other pickers up. In all innocence, I had shown them up.

If you have a young dog out for the first time or two, be careful what you ask him to pick up. Collect all the birds on bare ground while you leave him sitting then send him to find a bird the location of which you know exactly. Two or three birds at each stand is ample for the first time or two. Later on, if your puppy is slow at retrieving, you will find that this quickens him up enormously if he goes for several retrieves, one after the other. Never send a very young dog after a runner until it is out of sight. If you let him go while he sees it running, it is certain he will run in the next time. Don't have more than one pup out working at a time. They are sure to gang up on you!

If you are asked to beat through a wood let your older dog work, but have your young dogs walking to heel. You will find, if you are a spaniel handler, the beaters will go too fast for your young dogs who have no time to investigate a scent or bush or drop to shot.

A picking up dog soon becomes a really good marker and it is nothing for your dog to mark several hundred yards away, whilst he should be capable of being handled out to a bird at that distance. This is just as easy for springers as labradors with practice. Some older dogs learn when a bird is pricked, although it is not apparent as it passes overhead. I have three such dogs. They just give me a look as if to say 'Hurry, let me go' and off they tear; they may be away for fifteen minutes but invariably they are right, though I could swear there was nothing wrong with the bird.

Remember to study the wind carefully when walking to your position, i.e. don't go past a piece of cover to be shot later, however silently, if the wind blows from you to the cover. I have mentioned this before but it is so important it is worth mentioning again. All wild animals have an acute sense of sound and smell, and send out their danger signals which are passed on to each other. If you cannot tell from which direction the wind is blowing, lick your finger, hold it above your head and you will soon discover that one side is colder than the other. This is the direction from which the wind is coming. Be sure to take your dog's collar off before letting it hunt out of sight. The metal in it could give him a nasty shock if he touched an electric fence. He could also easily get caught up on a stake or blackthorn and strangle to death. Beware sending a puppy for a strong-running cock. Spurs can damage and frighten the puppy. Wait, also, until a hare is quite dead before sending him. I had one dog who had his gum slit open by a kicking hare, and he was always windy of picking up fur again.

You may find that your puppy does not like the taste of woodcock or snipe. This can be helped along at home by shooting starlings and moorhens for him. They are bitter, like woodcock, and if he once gets used to picking them up, one usually has no trouble with him on the real thing later on.

If you have an accident, go at once to the keeper and explain—don't let him hear of your misdeeds from someone else. There was the time time I was picking up with Jim Wylie near Thetford, where there were several preserved golden pheasants near the big house. I had my reliable old dog David with me. Normally, he never pegged, i.e. caught a pheasant unshot, in a bush, (or blinked, left his retrieve), but this day the golden pheasants were driving him mad. He would try and push them out of the bush and they would wait until the last minute and get up, enjoying wiping their long yellow tail feathers over his face. At last he could bear it no longer and snapped one up in a bush, there was a quick crunch, he scraped some leaves over it, and came out looking innocent. I was very upset, but Jim Wylie said he would take it to the keeper who told him to dispose of it. He wanted to take it to a friend who tied his own flies for fishing, so he put it in the partition under his seat. Five or six weeks later, we met again at the same place and I took my lunch to eat with Jim in his car. 'Phew, Jim . . .!' 'Yes, by jove, come to think of it, there *is* a smell. Wonder what it is?' We made a full search of the car and found nothing, until suddenly we saw maggots crawling out from underneath the seat. He had completely forgotten the golden pheasant!

My spaniel, Busy, taught herself a very useful trick. If she finds a runner that pops up into a bush or branch and she cannot get it down she will tear back to me and lead me, barking, to the spot.

If you do your job thoroughly, you won't be in for the drinks before

90. The finished article.
Gosling

lunch or available for the goodbyes at night. This is when you should be looking for long-range birds. When you have cleared up your last bird of the day, report to the headkeeper any ground not covered, or any known unpicked birds to be collected the next day. If a particular shoot has sufficient pickers-up on the shooting day, you may be only too welcome to come the following day to pick up cold. Here you do not need such a controlled dog, although some shoots don't like their birds harried two days running.

At the end of the day, don't stuff your dog in the back of the car like an old pair of gumboots and forget about him. Respect his feelings. Dry him off if soaking wet, with a chamois leather, wiping the same way has his coat lies; offer him a drink and give him a couple of biscuits. Sometimes, if it is very cold, I share my thermos of soup and sandwiches with the dogs. I know I feel better for something solid inside me and I am sure they do, too. When I get home there is a shed with an infra-red lamp and clean straw to dry them off. They then come in for inspection for thorns in their pads and to see their eyes are unscathed, etc. They have worked hard all day, and given their best, and the least you can do is look after them properly when they get home.

You will come off the shoot tired, muddy, stung by nettles, with wet, bedraggled dogs at your heels, their ears full of burrs; you will perhaps have twenty miles to drive home but will have an inner contentment in having done a good day's work and saved wounded birds from a lingering death.

List of Registered Breed Societies

SETTERS

Belfast and District Irish Setter Club—*Mr M. C. Lockley, 14 Beechill Park East, Belfast 8, N. Ireland.*

British Gordon Setter Club—*Mrs L. Jackson-Richmond, Garden Cottage, Sutton Park, Guildford, Surrey.*

English Setter Association—*Mrs V. Foss, 'Greenhill', 8 Reservoir Road, Woolton, Liverpool 25.*

English Setter Club—*Dr. J. B. Maurice, Batts Farm, Wilton, Marlborough, Wilts.*

Northern English Setter Society—*Mr S. J. Crane, 29 Woodside, Witton Park, Bishop Auckland, Co. Durham.*

Gordon Setter Association—*Mrs M. I. Smith, Riddell Mill, Lilliesleaf, Melrose, Roxburghshire, Scotland.*

Irish Setter Association (England)—*Mr W. J. Rasbridge, 105 Manor Green Road, Epsom, Surrey.*

Irish Setter Breeders Club—*Mrs A. I. Ashton, Greystones House, Horbury Road, Ossett, Yorks.*

Irish Setter Club of Scotland—*Mrs E. A. Groundwater, 20 Drum Brae South, Edinburgh 12.*

Ulster Irish Red Setter Club—*Mr C. Campbell, 42 Rochester Avenue, Belfast BT6 9JW.*

POINTERS

Northern Ireland Pointer Club—*Mr W. R. Gardiner, 12 Swanston Crescent, Glengormley, Newtownabby, Co. Antrim.*

Pointer Club—*Mrs S. M. Rhodes, 'Hazelbarrow', Walseker Lane, Woodhall-in-Harthill, Sheffield, Yorks.*

Southern Pointer Club—*Mrs G. Woollard, Willow Cottage, East Grimstead, near Salisbury, Wilts.*

GERMAN SHORTHAIRED POINTERS

German Shorthaired Pointer Club—*Mr W. K. Simpson, Northwold House, Northwold, near Thetford, Norfolk.*

HUNGARIAN VIZSLAS

Hungarian Vizsla Club—*Mrs M. Foster, Heather Cottage, Sedling, Wearhead, Bishop Auckland, Co. Durham.*

Hungarian Vizsla Society—*Mrs K. J. Auchterlonie, Academy House, Crieff, Perthshire.*

RETRIEVERS

Curly Retriever Club—*Mr J. F. Spooner, 104 Lovely Lane, Warrington, Lancs.*

East Anglian Labrador Retriever Club—*Mrs L. G. Kinsella, 'The Mount', Fingringhoe, near Colchester, Essex.*

Flat-Coated Retriever Society—*Hon. Mrs D. Jessel, Collyers, Steep, Petersfield, Hants.*

Golden Retriever Club—*Mrs B. Webb, 116 Cambridge Road, Great Shelford, Cambridge.*

Golden Retriever Club of Scotland—*Mr D. McVicar, c/o Royal Bank of Scotland, 339 Byres Road, Glasgow, W2.*

Golden Retriever Club of Wales—*Mrs F. E. Allan, 36G Gelli Road, Pemberton, Llanelli, Carms.*

Labrador Club of Scotland—*Mr J. Kerr, 16 Braemar Crescent, Carluke, Lanarkshire.*

Labrador Retriever Club—*Lt. Cdr P. A. Whitehead, Brookmead, East Grimstead, Salisbury, Wiltshire.*

Labrador Retriever Club of Northern Ireland—*Mr A. J. Kilpatrick, 16 Corby Drive, Lisburn, Co. Antrim.*

Midland Counties Labrador Retriever Club—*Mrs J. Taylor, Chase Cliffe, Whatstandwell, near Matlock, Derbys DE4 5EQ.*

Northern Golden Retriever Association—*Mrs M. Dawson, The Poplars, Donington Northorpe, Spalding, Lincs.*

Northumberland and Durham Labrador Retriever Club—*Miss E. Smith, Brooklands, 29 High Street, Gosforth, Newcastle-on-Tyne NE3 4AD.*

North West Labrador Retriever Club—*Mr J. D. Lilley, The Firs, Parkfield Road, Knutsford, Cheshire.*

Three Ridings Labrador Club—*Mrs R. M. Chadwick, Northcliffe, Northcliffe Lane, Thornton, Bradford, Yorks.*

Ulster Golden Retriever Club—*Mr J. B. McKeown, 8 Glenshesk Park, Dunmurry, Belfast.*

West of England Labrador Retriever Club—*Mr L. J. Harding, Coombe Cross, Dittisham, near Dartmouth, Devon.*

Yellow Labrador Club—*Mrs A. Wormald, Glenstuart, Cummertrees, Annan, Dumfriesshire.*

SPANIELS

American Cocker Spaniel Club of Great Britain—*Mrs A. M. Jones, The Captains, Castle Morton, Malvern, Worcester.*

Antrim and Down English Springer Spaniel Club—*Mr J. S. Burrows, 92 Locksley Park, Finaghy, Belfast BT10 0AT.*

Black Cocker Spaniel Society—*Mrs E. M. Cox, Ragmans Castle, Handycross, Marlow, Bucks.*

Cheshire Cocker Spaniel Club—*Mrs N. Curtis, Horfield, Lostock Avenue, Poynton, Cheshire.*

Clumber Spaniel Club—*Mrs K. Vincent, The Steadings, Alveston, near Bristol, Glos.*

Cocker Spaniel Club—*Mrs E. Robertson, 'Nostrebor', Pikes Pool Lane, Finstall, Bromsgrove, Worcs.*

Cocker Spaniel Club of Lancashire—*Mrs D. M. Schofield, Cobbles, Norcott Brook, By Warrington, Lancs.*

Cocker Spaniel Club of Scotland—*Mrs J. Braid, Craigallian, Dunure Road, Ayr.*

Coventry Cocker Spaniel Club—*Mrs D. N. Chapman, New Fullbrook, West Bromwich Road, Walsall, Staffs.*

East Anglian Cocker Spaniel Society—*Mrs M. Bridgford, Hill Farmhouse, Drinlestone, Bury St Edmunds, Suffolk.*

East of Scotland Cocker Spaniel Club—*Mrs E. Johnston, Knock Tower, Newbridge, Midlothian.*

English Springer Spaniel Club—*Mrs L. B. Hampton, Larkstoke Cottage, Bourton-on-the-Hill, Moreton-in-Marsh, Glos.*

English Springer Spaniel Club of Northern Ireland—*Mr R. E. Clemitson, Fairacres, 1 Holywood Road, Newtownards, Co. Down, N. Ireland.*

English Springer Spaniel Club of Scotland—*Mrs E. K. Thomson, Rivington Lodge, Castle Douglas, Scotland.*

English Springer Spaniel Club of Wales—*Mr P. A. Davies, Springfields, Blackton Lane, Penmark, Barry, Glam.*

Field Spaniel Society—*Mr R. Hall Jones, The Captains, Castle Morton, Malvern, Worcs.*

Home Counties Cocker Club—*Mrs E. Samuels, Jolly Waggoners, Much Hadham, Herts.*

Irish Water Spaniel Association—*Mr N. Waters, Zanfi, Drury Lane, Somerwood, Rodington, near Shrewsbury, Salop.*

London and Home Counties English Springer Spaniel Society—*Mr D. Miller, 283 Desborough Avenue, High Wycombe, Bucks.*

London Cocker Spaniel Society—*Mrs K. Baldwin, 79 Hereford Road, Bayswater, W2.*

Midland Cocker Spaniel Club—*Mrs J. Caddy, Park Head House, Haughton, near Stafford.*

Midland English Springer Spaniel Society—*Mrs M. Backhouse, 121 Silcoates Lane, Wrenthorpe, Wakefield, Yorks.*

Northern Counties American Cocker Spaniel Club—*Mrs M. F. Slapp, 2 The Crescent, Broadwater, Fleetwood, Lancs.*

North Midland and Eastern Counties Cocker Spaniel Club—*Mrs W. M. Prince, Bowbridge Fields Farm, Mackworth, Derbys.*

North of England Cocker Spaniel Association—*Mr D. Sionister, 284 Stretford Road, Urmston, Manchester, M31 1WJ.*

North of Ireland Cocker Spaniel Club—*Mrs D. Yeames, 74 Bawnmore Road, Belfast BT9 6LD.*

North Wales Cocker Spaniel Club—*Mr G. I. Jones, Woodville, Woodlands, Conway, Caerns, Wales.*

Parti-Coloured Cocker Spaniel Club—*Mrs M. Phillips, Shielwood Kennels, Rossway Farm, Little Bushey Lane, Bushey, Herts.*

Red and Golden Cocker Spaniel Club—*Mrs E. M. Pain, Clough House, Parson Drove, Wisbech, Cambs.*

Rotherham and District Cocker Spaniel Club—*Mrs M. E. Stevens, Leabank Kennels, Fulshaw Cross, near Thurlstone, Sheffield.*

South Wales and Mon. Cocker Club—*Mrs T. M. Bebb, Bona Vista, 148 Cefn Road, Rogerstone, Mon.*

Sussex Spaniel Association—*Mrs Lancaster, Moss Cottage (Nursery Road) off Close Lane, Alsager, Stoke-on-Trent.*

Ulster Cocker Spaniel Club—*Mr T. J. Cardy, Mount Keepe, Glen Road, Lower Castlereagh, Belfast 5.*

Welsh Springer Spaniel Club—*Mrs A. Walton, The Lodge, Fairseat Lane, Wrotham, Sevenoaks, Kent.*

West of England Cocker Spaniel Club—*Mr T. E. W. Heavisides, 'Kirkdon', 26 Moorland Road, Yate, near Bristol.*

Yorkshire Cocker Spaniel Club—*Mr R. Middleton, Rimaelia Kennels, 20 Woodside Road, Silsden, Keighley, Yorks.*

WEIMARANERS

Weimaraner Club of Great Britain—*Mrs M. Holmes, 78 The Byeway, West Wittering, Sussex.*